A Life
of Ospreys

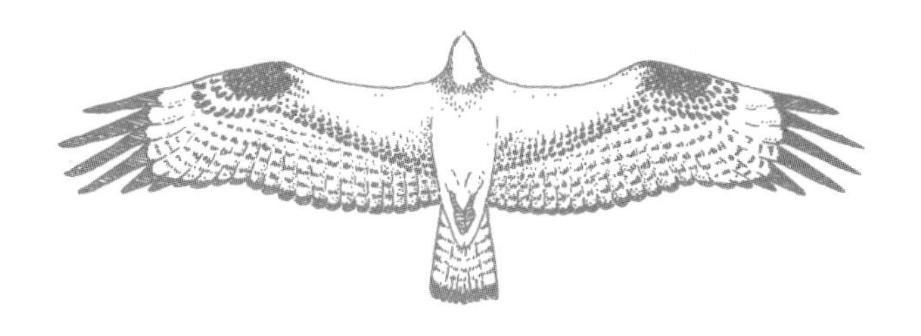

A LIFE OF OSPREYS

Roy Dennis

Whittles Publishing

Published by
Whittles Publishing,
Dunbeath,
Caithness KW6 6EG,
Scotland, UK
www.whittlespublishing.com

© 2008 Roy Dennis
ISBN 978-1904445-26-5

Reprinted 2012

Drawing on cover, and sketches throughout the book by Alessandro Troisi, Rome

All rights reserved.
No part of this publication may be reproduced,
stored in a retrieval system, or transmitted,
in any form or by any means, electronic,
mechanical, recording or otherwise
without prior permission of the publishers.

Typesetting and layout by Mark Mechan

Printed in India by Imprint Digital

In memory of George Waterston, OBE (1911–1980), who set me off on a lifelong association with ospreys, with huge enthusiasm and encouragement

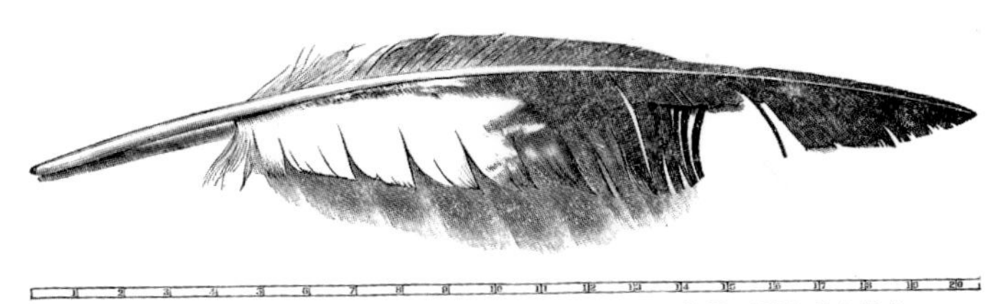

OSPREY'S WING FEATHER (1st primary) picked up on 16th May 1892 by J. A. H.-B., below the tree where the new nest was.—*From Photo by* Mr. Brown *of Falkirk.*

Contents

Acknowledgements

As this book will show, those who most merit sincere thanks for their work with ospreys and for kind permission to work on their lands cannot be named: landowners, farmers, keepers, foresters and members of the public who have kept watch over individual nests while keeping their location a well-guarded secret. For continuing, even today, to respect 'their' ospreys' privacy, they are owed an enormous thank you – even if it must still, for the sake of the birds, be anonymous.

I would like to thank those colleagues in Scotland and elsewhere in the world who have shared my passion for ospreys. Their knowledge, hard work and vision are helping to restore the osprey population to where it should be. And they, like me, have never lost sight of the fact that working with ospreys should always be exciting, inspiring, and, quite simply, great fun.

I thank the photographers and artists who have given me images over the years, some of which appear in this book, hopefully acknowledged – apologies to any I have missed.

I am exceedingly grateful to many people for practical help in the field, information, advice and other kindnesses – many of them are mentioned within the book. I am indebted to the RSPB, especially the membership who provided my wages, for whom I worked for over two decades; and in the last twenty years the private individuals and companies who have supported our osprey work through the Highland Foundation for Wildlife. Anglia Water plc generously funded the early satellite studies and recently Talisman Energy (UK) Ltd have been very kind in funding the new GPS satellite tracking project. The Forestry Commission in Scotland has been a generous partner through all the years, while Scottish Natural Heritage (previously the Nature Conservancy) and the British Trust for Ornithology have licensed my many conservation activities. Sincere thanks to all.

Preface

I saw my first osprey 48 years ago at Loch Garten, a secluded corner of the Scottish Highlands now synonymous with these magnificent birds. Back then it was just starting a long association with the return of the osprey to Britain, and so, it turned out, was I. Nineteen years old, a new warden at the osprey camp run by the Royal Society for the Protection of Birds (RSPB), I was watching, half excited, half anxious, for the return of Scotland's only breeding pair. The first week of April had gone by and not a sign. On 8 April 1960, a cold grey wet day, my early morning visit to the eyrie had again drawn a blank, and as I trudged back hunched against the rain for my afternoon check, I had no real reason to be optimistic. I looked through the drizzle across the forest bog and there – I can see it still – in the branches of an ancient Scots pine, was the male, peacefully preening his wet feathers, his three and a half thousand mile migration from Africa behind him, and back in Scotland for another season. I waited and watched him while he rested, recording over an hour his every move until, as the afternoon drew to a close, he flew off to fish.

More waiting and watching was to come, ten whole days of it, until the female arrived to join her mate. At least I got to know him well.

This was the first of four life-changing summers spent helping to protect these ospreys. Like so many people before and since, I fell under the spell, not only of these beautiful birds but also of Strathspey, its wildlife and its people. In 1971 I came back to work in the Scottish Highlands and have studied ospreys at home and abroad ever since. They have given me such an exciting life journey – a tale of successes and failures. Now, as I write, I have the immense satisfaction of knowing that ospreys are breeding in England and Wales, and that the Scottish population has risen to nearly 200 pairs – the highest for more than two centuries, and a success unimaginable in those pioneering days of 'Operation Osprey'.

This book endeavours to bring the osprey story up to date. It covers the birds' tragic past, those early days at Loch Garten and the gradual process of recolonisation. We no longer have to wait to see an osprey return – satellite technology, undreamt of in those early days, allows us to follow some of them on migration, and the data from transmitters is included here. There are tales of favourite birds such as 'red Z', a breeding female which I rescued from certain death and who went on to live to the age of 23, rearing 32 young. And, of course, there's Henry and EJ, the latest residents at Loch Garten, reality TV stars in their own right, their every move followed by thousands via cameras in the nest.

The thrill of seeing that first osprey, back in 1960, has never left me and is reawakened every spring when I see the birds back on their nests, safely home from migration. However many ospreys I have seen across the world in nearly half a century, the discovery of a new nest or the sighting of a favourite bird is as exciting as ever; the piercing yellow eyes returning my gaze through the telescope bringing the promise of another successful season. I hope that through this book, you too will come to experience some of this excitement or rekindle your own memories of the 'fish hawk'. May you enjoy ospreys as much as I have done through the years, for they are a glimpse of truly wild nature in our increasingly circumscribed and industrialised world.

Roy Dennis MBE
The Scottish Highlands

1
Introduction

The osprey is one of the world's most iconic and well known birds. It ranges throughout all five occupied continents, and wherever it is found, its reputation as a peerless and spectacular fisherman remains unchallenged. It is a bird totemic of its environment, the hounding of past years replaced by widespread respect and admiration, and if left unpersecuted, the osprey is happy to live within close proximity to man.

The osprey is a highly distinctive bird of prey, the only raptor that catches nothing but fish for its food. It is relatively large, growing to about 55–62 centimetres in length (approximately two feet) with exceptionally long wings that extend to a span of 145–180 centimetres (over five feet). These wings are often held in a characteristic 'W' shape, making the bird unmistakeable in flight. When compared with other birds of prey in Britain, it appears larger than a buzzard but obviously smaller than a golden eagle, and is usually associated with inland or coastal waters. In colour, the osprey is a two-tone bird, being dark brown above and white below and, in Britain, can occasionally be confused with some of our larger gulls. Ospreys weigh between 1.2 and 2 kilograms (approximately two-and-a-half to four-and-a-half pounds) with males some 5–10% smaller than females. Northern ospreys are larger than those in southern latitudes.

The osprey's upper body is dark brown while the underparts are white, with a variable brown band running across the breast. The head is white with a dark brown band, or 'eye stripe', running across the eye to the back of the neck. Both male and female have a slight crest on the

Osprey face – with its characteristic brown 'eye stripe' and slight crest

top of their head. There is brown on the osprey's breast, the female being much more heavily marked. The short tail is barred brown and buff, appearing paler at the edges in male ospreys. The under-wing is white, especially so in the male, and contrasts clearly with the dark brown flight feathers. The eyes are yellow (although orange in the young) and the voice is capable of a range of shrill calls and whistles. The legs, greenish-grey in colour, are remarkable for their stout and powerful long black talons. No one who sees an osprey dive for fish, though, will need the following prosaic checklist: that unmistakeable plunge, the fierce grip of talons on a fish ripped from water to air, the characteristic shaking of water from the feathers. Such traits could only ever belong to an osprey.

The osprey's majestic fishing abilities are thanks to highly specialised features, enabling this bird so graceful in flight to remain equally adept in water. The strong feet are ideally suited for grabbing a fish, the outer toe being reversible, meaning that prey can be gripped firmly between two talons in front and two behind. The skin of the foot is prickly and coarse, allowing a secure hold on slippery fish, and the strong black bill is curved into a hook for tearing up prey. The bird has a fleshy nostril that closes as it dives into water and its feathers, particularly those on the under-parts, have their own adaptation: they do not possess an aftershaft. This is a tiny extra feather affixed, in most birds, to the base of larger feathers. Its absence increases the osprey's ability to shed droplets once it has left the water. Ospreys also have a large preen gland and spend much of their time re-oiling their feathers, a process crucial to a bird that relies on being waterproof for its survival.

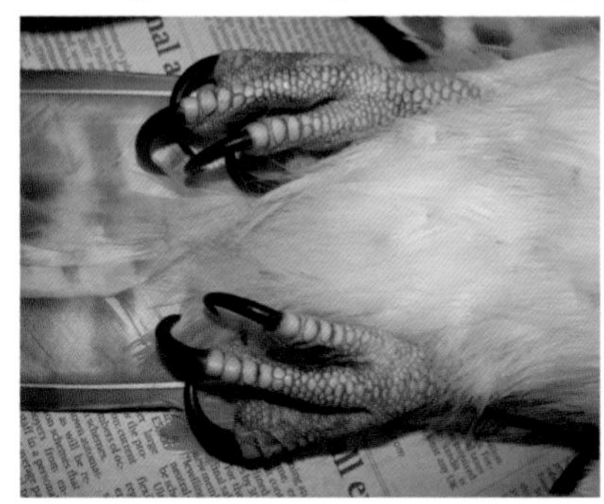

Feet and talons

It is thought that there are around 40,000–50,000 pairs of ospreys in the world. In Europe and North America the numbers are increasing but elsewhere, especially in east and southeast Asia, the osprey appears to be in decline. The bird has a worldwide distribution – the peregrine being the only other diurnal raptor with a similarly wide range – and, perhaps surprisingly, there is not just only one genus but also only one species, divided into a number of different subspecies or races.

The scientific name of the osprey is *Pandion haliaetus, Pandion* coming from the name of a king of Athens, while *haliaetus* is derived from the Greek *hals*, meaning 'salt' or 'sea', and *aetus*, meaning 'eagle'. It was first scientifically described as *Falco haliaetus* in 1758, by the Swedish scientist, Carl Linnaeus (1707–1778), who mistakenly placed the osprey in the same family as the falcon.

Later, in 1809, Marie Jules César Savigny (1777–1851), named the modern genus *Pandion* but he also changed the species name to *fluviatilis*. Finally, René Primeverè Lesson (1794–1849), in 1828, combined the names as *Pandion haliaetus* and so it has remained to this day.

The name 'osprey' comes from the Middle English *ospray*, derived from *ossifraga* – the 'bone-breaker'. This in itself is a misnomer since it refers to the lammergeyer, or bearded vulture. The French name, *balbuzard*, is thought to have originated with British settlers, who saw ospreys and called them 'bald buzzards', the word later becoming corrupted into its present form.

The familiarity of the osprey to so many different peoples has resulted in ospreys having many common names. In England they used to be called 'fish-hawks' or 'mullet hawks', while in Scotland the Gaelic name for the osprey, *Iasgair*, means 'fisherman'. There are four subspecies of ospreys in existence, although some people recognised even more in the past. Northern ospreys are highly migratory while those in mid-latitudes and Australasia are mainly resident or nomadic.

In the geological record the line of the osprey stretches back 50 million years, with one long-extinct species, *Pandion homalopteron*, identified in California, dating back approximately 13 million years. Other fossils suggest that ospreys were widespread some 10–15 million years ago and it would seem that these birds were pretty similar to those of the present day. Approximately one million years ago ospreys were well distributed throughout Western Europe and North America. There seems to be no doubt that ospreys resembling those known to us were living in Europe when the earliest of our ancient ancestors arrived from Africa.

The first appearance of the osprey in literature occurs in Aristophanes' *The Birds*, written in Athens in 410 BC. The osprey is mentioned only in passing, although the variety of birds in the play (the main character is a hoopoe) suggests that Aristophanes may be the earliest documented bird watcher. Sixty years later Aristotle wrote in his *Natural History*: 'There is another species, called the sea-eagle or osprey. This bird has a large, thick neck, curved wings, and broad tail feathers; it lives near the sea, grasps its prey with its talons, and often, from inability to carry it, tumbles down into the water.'

Male osprey preening on roost – the osprey's feathers and preen gland are adapted to ensure it can shed water quickly and remain waterproof

Whatever Aristotle may have thought of its abilities, the osprey is a highly adaptable bird, occupying salt, brackish and freshwater habitats throughout the world, and catching the most amazing variety of fish species – whatever, in fact, is available. It will also use whatever presents itself as nest material, building large nests, mostly from sticks, in almost every conceivable locality; in trees, on the ground, in cliffs and even on man-made structures. I have seen ospreys nesting on the ground on the coral reef islands of the Arabian Gulf; on extinct volcanoes on the Cape Verde islands; on spiky sea-cliff pinnacles of Corsica and Hokkaido, Japan; on electricity pylons in eastern Germany and in Queensland, Australia; in ancient Scots pines in Scotland and Sweden, and in man-made nests and on navigation markers in North America. In fact, they will nest anywhere on anything, as long as they are free from ground predators.

The author with George Waterston, monitoring seabirds by boat, Fair Isle, 1968 (photo courtesy of J. Arnott)

History of the osprey watcher

In 1959, the year before my association with ospreys began, I spent a year as the assistant warden at the bird observatory on Fair Isle, a wonderfully remote and scenic island lying midway between Orkney and Shetland, off the north coast of mainland Scotland. In September, as the autumn migration started, a special visitor arrived: George Waterston, the Scottish director of the Royal Society for the Protection of Birds (RSPB). The observatory had been his brainchild and he was both its secretary and its driving force. Meeting such a great ornithologist was to change my life. One evening George started to tell us about the excitement of the summer just past, watching over the successful nesting venture of a pair of ospreys at Loch Garten. That one evening was enough: I determined to discover more about these great raptors.

When I left Fair Isle in November of that year, George and I discussed the possibility of my spending the next summer working at Loch Garten and, in the New Year, I duly applied for the post of osprey warden with the RSPB. I still remember how excited I was to get the job, and on the first day of April, 1960, I travelled north with George to the place that was to play such a central role in my life. There was so much to see and learn in this most beautiful part of the world and Strathspey soon became my home. I count myself very fortunate to have known

15th February 1960

Dear Roy,

Operation Osprey 1960

I am delighted to hear that you will be available as a paid warden and I write at once to say that we would be very glad indeed if you would accept the post. ...We could then set off by car, Friday 1st April, to establish the Base Camp and begin operations at Boat of Garten. You would be engaged as Junior Warden as I feel that we shall have to get someone older as Senior as the duties would involve taking full charge of Operation Osprey and ordering much older people than yourself (Group Captains RAF etc.!) about. As soon as the appointment of Senior Warden is made I shall let you know.

All the best,

Yours ever

George

Letter that started my life's work with ospreys: from George Waterston, Scottish Director RSPB

George; he was one of the most influential figures in my life. Both he and his wife, Irene, were generous with their time and experience, introducing me to people and to places throughout Scotland and giving me much sound advice over the years. Sadly, George suffered from kidney failure and died long before his time. I often wish that he could know how well his ospreys have flourished and that his project to reintroduce the white-tailed sea eagle has met with equal success.

For four happy and exciting summers, I worked with Dick Fursman, a recently retired RAF Wing Commander who had become senior warden. He and his wife, Marian, became great friends and, nearly fifty years on, he now lies buried at Insh, where a pair of ospreys breed overlooking the tiny churchyard. I always think of him when I check that nest, and of how the story of the ospreys' return is, in great part, the story of fine people like him.

Operation Osprey's camp, 1978

During those early days, watching as the birds raised their young, we welcomed thousands of people to the public observation post at Loch Garten, and were visited by famous ornithologists from every corner of the world. We worked with the most fantastically loyal group of volunteers, who returned year after year to the 'osprey camp' hidden in the forest on a Miss MacDonald's croft. She was one of the greatest supporters of George and his ospreys and helped us with moral and practical support.

The camp was a very basic place. We all slept in tents, although there was a caravan used by the cooks who, somehow, managed to produce marvellous meals. In our first year we put up a wooden shed, ceremoniously named 'Pandion Palace' and used for off-duty times. George had devised rotas, so that Operation Osprey ran like clockwork, based on three teams of volunteers covering the entire day and night. Many volunteers were regulars, returning year after year.

Those early days were such great fun and we knew we were doing something worthwhile. Again, the osprey story becomes a human one, with help coming from many great friends whose own stories were interwoven with that of the birds. There were local Strathspey people like Colonel Iain Grant of Rothiemurchus, whose ancestors had been fierce defenders of the ospreys at Loch an Eilean in Victorian times, Willie Marshall in Nethybridge who was the Chief Forester in that area for the Countess of Seafield, one of the leading landowners at the time and the owner of the osprey sanctuary, and Willie's nephew Hamish, along with Alistair

Successful season

25th August 1960 from George Waterston:
Letter to all wardens and volunteers thanking them for the season's help and informing them of the year's successful fledging of two young. He adds: 'After the young had left, Roy climbed up to the nest with extreme difficulty – aided by a ladder which he had to drag up into the fork of the tree and lash in position. He retrieved many fish remains among which were many jawbones of pike'.

McCook, who helped us with the evening watches. The project seemed to capture the imagination of the local people, with Highland businessmen, hotel keepers and customs officers helping us regularly with watches alongside the annual volunteers, many of whom were senior Army and RAF officers who treated the project as a kind of campaign. George, after all, ran the place with military precision.

In 1963 we suffered a set-back. Storm-force winds tore down the nest at Loch Garten, but we simply upped sticks and moved our operation to Loch an Eilean, just a few miles away. There we set up a second camp in a cottage on Colonel Grant's estate, allowing us to protect the nest of a new pair of ospreys on Forestry Commission land at Inshriach. Sadly, the eggs failed to hatch that year, but still it was good news. The new pair meant osprey numbers were increasing and the future for them was beginning to look brighter.

That autumn, I returned to Fair Isle to run the bird observatory. This time I was not alone, being newly married to Marina, Miss MacDonald's niece, whom I had met three years before at the croft. We spent seven great years on the island, where seabirds and bird migration were the focus of my work. Still, I kept up to date with the ospreys' progress, meeting up with friends from Operation Osprey during the winter months. And, very occasionally, I would see the odd lost migrant osprey passing by Fair Isle, imagining it looking down on my home island, just three miles by one, a rocky outcrop in the cold expanse of the North Sea.

In 1970 we said goodbye to Fair Isle and moved back to the mainland with our three children, Rona, Gavin and Roddy. I had been appointed as the new Highland Officer of the RSPB and my life with birds was about to take a different turn.

By 1971 there were seven pairs of ospreys nesting in Scotland – in Badenoch and Strathspey, Moray and Perthshire. Their conservation and protection was part of my RSPB conservation work, and I also found myself responsible for Operation Osprey, which was continuing at Loch Garten. A succession of Loch Garten wardens, including Harvey Burton, Russell Leavett and Stewart Taylor, helped with the work at other nests too, everything from climbing trees to keeping watch around the clock in an effort to protect the birds. Soon, I was ringing my first young ospreys, following on from Doug Weir who had begun the work in 1966 – and every year, to this day, I ring as many young birds as I can, with help from friends and colleagues.

Ringing osprey from nest B09, 1976 (photo courtesy of A. Laing)

inflation is essential.
—R. E. Dundas.
N Scot 1/7/72

Osprey 'Special' Brings 500 To Speyside

NEARLY 500 English wild bird enthusiasts arrived at Aviemore station by train from Carlisle on Saturday.

The bulk of the party travelled in buses to Loch Garten to see the ospreys, giving the R.S.P.B. viewing post its biggest single tourist party since the birds returned to the loch in 1964.

They were met by Mr Roy Dennis, Aviemore, the Highland Officer of the Society.

But again, this story comes back to people. One of the greatest satisfactions of my conservation work with ospreys has been meeting the gamekeepers, foresters, farmers, crofters, factors and landowners who had birds nesting on their land. Many friendships were established, and, as the exact location of the nests had to be kept a secret, confidences were scrupulously kept. The friendships I made then were to survive over decades, and I often used to think how well the ospreys chose the protectors of their nest sites. These friendships and contacts were handed down through generations and, today, I may find myself dealing with the second or even third generation to have an osprey nest on their land. In 1972 I started to produce an osprey newsletter, using code numbers for each nest described to ensure security. There were, and are, so many people I would like to thank but in order to safeguard nests and nesting sites, it is still essential to maintain confidentiality.

In the 1970s we started the practice of rebuilding damaged nests and began to try to extend the population by building artificial nests in new areas. Both these ventures soon yielded results and greatly helped us increase the production of young. These were pioneering days for getting ospreys mentioned in the media, and a tiny foreshadowing of things to come. In 1977 the RSPB commissioned an ambitious film about the birds and the eminent wildlife cameraman, Hugh Miles, produced a landmark film *Osprey*, a favourite RSPB film to this day. During the filming, Hugh and I visited the Gambia to discover more about how the birds spent their winters so far away from Scotland.

In October 1980 I was fortunate to attend an international symposium on ospreys and bald eagles in Montréal; there I met many osprey workers from North America and also from Corsica and Finland, which considerably broadened my knowledge of ospreys worldwide. It was very encouraging to see how well ospreys had started to crawl back from the problems of pesticide contamination in North America after the banning of the most persistent chemicals. I also learnt that in some North American states ospreys were being reintroduced to their old haunts by the translocation of eggs and young.

If ever I need to remind myself of how far the story has come, I think of the day in 1983 when the RSPB celebrated the arrival of the millionth visitor at the Loch Garten centre. With each new advance in technology, further opportunities to spread the osprey message were created until, in 1987, the BBC set up an 'Osprey Watch' at a nest site in Moray. With a closed-circuit television camera positioned in the tree next to the eyrie and an outside broadcast crew, I presented twelve live broadcasts during the nesting cycle. It might seem commonplace now, but it paved the way for many other live outside broadcasts featuring our native wildlife.

By this time, the osprey population had increased greatly and osprey monitoring in other parts of Scotland had passed to local enthusiasts, particularly Keith Brockie in Perthshire, Roger Broad in Stirling and Argyll, and Iain Macleod and Keith Duncan in Aberdeenshire. I continued to share the monitoring of ospreys in the north of Scotland with Colin Crooke.

In 1991 I left the RSPB to become self-employed and, set free from office duties, I was able to spend more of my time during the summer studying and helping to conserve ospreys. It had always been a labour of love with me, involving early mornings and late evenings before and after my work. Now, though, I had time to identify and follow the progress of more of the colour-ringed birds and study how they fared within the total population. New techniques allowed us to radio track male birds in Strathspey one season and, later, to follow their journeys to Africa using satellite transmitters.

In 1996, along with Tim Appleton of the Rutland Water Nature Reserve and Stephen Boult of the water company, Anglian Water plc, I was able to start the first experimental translocation of ospreys within Europe, following in the footsteps of successful projects in the United States. This has led to similar projects in mainland Europe, with the first, in southern Spain, having a successful outcome in 2005, when the first pair bred in Andalucia. In 2007 I was thrilled to hear that a chick born in a nest at Rutland Water had returned there and reared two young herself. Another important step forward.

Interest in ospreys is now greater than ever before, with at least seven permanent visitor centres in the UK devoted to these intriguing birds, giving members of the public an opportunity to find out more about them. More and more people are enjoying the thrill of seeing an osprey, whether unexpectedly at a local lake, river, estuary or loch, during a planned visit to a nature reserve, or while abroad, where they appear like well known faces in an unfamiliar setting. Some ospreys have even become household names and although I have never been keen on naming wild creatures, I do recognise that the exploits of the latest occupants of the Loch Garten nest, Henry and EJ (named after the colour rings I put on their legs as chicks – HV was

Author weighing chick, 2007

first christened Henry the Fifth, soon shortened to Henry) are eagerly followed by thousands of people on the RSPB's osprey website. Widespread fascination with these birds can surely only lead to greater understanding and, therefore, to greater protection.

The author with Logie, a previously unringed adult female, having been caught for tracking at nest B10 in Moray, 2007. Logie was fitted with the very latest in GPS satellite transmitters and her incredible migratory journey has featured on BBC Radio 4's World on the Move: Great Animal Migrations *programme. (The author's diary of Logie's migration can be found at: www.roydennis.org)*

I still relish the challenge of trying to get ospreys back into their ancestral range throughout the British Isles, where their population should be closer to 2000 pairs than a mere 200. The thought that, one day, they will once again catch fish in the Thames in central London is not only a thrilling dream, but, now, a realistic one. It never fails to inspire me. On a wider front, I enjoy encouraging and helping colleagues in mainland Europe to carry out projects that will ensure the full return of the osprey to its southern range. I have been blessed, throughout my life, to have worked with such a fantastic bird and with so many committed and hard-working people and I look forward to many new encounters with this exciting raptor, both at home and abroad.

2

Famous places, special birds

Loch Garten

Think 'osprey' in Britain and it is Loch Garten that first springs to mind: it was here that the modern history of the bird in Scotland began. It is a wonderfully tranquil place, in Strathspey in the Highlands of Scotland, and for years I lived just near it. I have seen the loch completely ice-bound in wintertime, its surface a snowfield – a sight that might be consigned to history now, thanks to climate change. Enclosing the loch are the ancient dark green Scots pines of Abernethy Forest and behind them, beyond heather-clad slopes, is the gentle, smudgy profile of the Kincardine hills. Away to the south-east rises snow-capped Ben Bynack, while hidden in the south lie the Cairngorm mountains. Standing here, you might almost be in Scandinavia.

Loch Garten and Loch Mallachie are part of the RSPB's Abernethy Nature Reserve. Both lochs of outstanding beauty, they often appear calm with only the watery trails of black and white goldeneye ducks to ruffle the mirrored reflection of the great pines around the shoreline. This is a landscape formed by the great ice sheets during glacial times. Look carefully and you see that the waters are stained dark brown with peat, acidic and unproductive, sustaining very few fish and so not a favoured hunting ground for ospreys. No rivers run in or out of Loch Garten, the water simply seeping through the peat mosses studded with hillocks of pine.

In 1954, in those peat mosses at the south end of Loch Garten, the ornithologist, Desmond Nethersole-Thompson, recorded the presence of a pair of ospreys with two young. This was kept secret for many years but has become the recognised starting point of the new era, heralding the recovery of the osprey population in Scotland. In 1955 the birds returned to the same ancient Scots pine tree in the bog, but the secret had got out and their eggs were stolen. Later that season, far too late for rearing young, they were seen building a frustration nest in the Sluggan Pass some miles away. In 1956 eggs were stolen from an eyrie in Rothiemurchus forest and the osprey pair involved here moved and built a new frustration eyrie near Loch Morlich. By this time a small group of RSPB staff and volunteers, along with Colonel Iain Grant of Rothiemurchus, were trying to protect the birds but, despite their efforts, the ospreys continued to have their nests robbed of any eggs that they managed to lay.

The following year George Waterston, the director of the RSPB in Scotland, organised a more determined protection effort using a group of dedicated wardens, but only one osprey

Loch Garten

returned to the old eyrie at Loch Garten. Sadly, there were even rumours that another osprey had been shot elsewhere in Strathspey.

The year 1958 began on a much more hopeful note. The male was back at the Loch Garten nest on 1 May and was joined two days later by a female. George Waterston immediately started the RSPB's protection project that was to become known as Operation Osprey. He acted in the nick of time as, on 11 May, on the very day that the female laid her first egg, a known egg collector was seen at the tree and chased off. Even so, despite a round-the-clock watch on the nest by a team of wardens, an egg collector managed to raid it during the very dark and rainy night of 3 June. The two eggs were found smashed at the foot of the tree and two hen's eggs, daubed with brown shoe polish, were discovered in their place in the nest. The thief was never traced. It was a total disaster.

The pair of ospreys moved and built a frustration eyrie in a tall Scots pine growing on a hillock in the mossy area to the north of the loch. The next year, in 1959, with the kind co-operation of the Countess of Seafield, the owner of Abernethy Forest, the area around Loch Garten was declared a protected bird sanctuary by the Secretary of State for Scotland. From then on it would be an offence to enter the area near the nest without permission. That year too, the new Loch Garten nest was strictly protected and the RSPB organised in advance a full complement of wardens and volunteers to guard over the ospreys should they return to breed. The male arrived on 18 April and his mate on 22 April. Courtship and nest building got underway and the first egg was laid on 1 May. On 8 June 1959 the behaviour of the birds suggested that the eggs had hatched and, sure enough, for the first time there were young in the nest.

George Waterston was undoubtedly a visionary. There had been increasing interest in what was happening at Loch Garten and it was proving very difficult to keep the location of the ospreys a secret. Rather than carry on trying, he went the other way, deciding to publish the great news and to invite members of the public to an observation post for a view of the ospreys and their young family. In the seven weeks until the young flew, 14,000 people came to see the birds. George's far-sighted decision proved to be one of the most decisive turning points in the conservation of ospreys and, later, of other rare birds. From that early RSPB observation post, Loch Garten went on to become one of the most successful and enduring visitor centres and set an early example in what we now refer to as eco-tourism.

THE EYRIE PROJECT

THE ROYAL Society for the Protection of Birds are to be congratulated on their most recent initiative—and it is to be hoped that the public will support them not only by dipping into their pockets to help out but by visiting the famous osprey eyrie at Loch Garten which they have bought. For the £290,000 purchase from Seafield Estates there is not just the site of "Britain's best known bird's nest," but 1500 acres of surrounding pinewood, moorland, loch and marsh, taking in a substantial part of the old Caledonian Forest.

The cost is high mainly because of the value of the timber, which by careful management will become an even better habitat for wildlife—and the society feel it is unlikely they will get another chance to buy such an outstanding area. Without doubt it is an exciting development, since this rare forest heritage can be truly protected for future generations to enjoy. Mr Frank Hamilton, deputy director of the RSPB, believes that the Speyside land deal may be the largest and most important by a single UK voluntary organisation and possibly even in Europe.

From the point of view of the Highlands, this constitutes not only wise action to protect the environment and threatened wildlife but the consolidation of an important tourist feature. It is best that all our eggs should not be in one basket —or nest. Diversification and determination to protect natural amenity will together ensure that the area's attractions are not only preserved but enhanced.

Efforts must be continued to safeguard our heritage. There are times when a progressive community should be prepared to invest in the past to enrich its future.

The Press and Journal *report of the Eyrie Project, 12 November 1975 – praise for an early example of Loch Garten's progressive conservationism and 'ecotourism'*

My first osprey

Friday, 8th April 1960. After lunch I went through the forest to the hide and had a wonderful surprise when I found an osprey sitting in a dead tree behind the eyrie tree. A really wonderful bird with indistinct brown pectoral band, crest, white underparts and dark brown upper parts. From 1432 to 1500 it just sat in the tree; at 1535 it brought a stick to the nest, then it brought in another six sticks and rearranged the nest. It left for the fishing lochs at 1550. I telephoned George and Dick with the news.

The Loch Garten eyrie was built on a great venerable Scots pine with two large branches spreading from the trunk about half way up and a broad flat top, a fine example of what are called 'granny pines' that must have started growing at least as far back as the middle of the seventeenth century. The ospreys built their huge eyrie on the canopy, covering much of the top. It was one of a group of three ancient trees: directly to the south of it stood an even older dead tree, the branches of which provided perfect perches for the ospreys, while on the other side grew another smaller live pine.

RSPB observation post, Loch Garten, 1970

When the first hide was opened at Loch Garten, access was via a small peat track leading for a few hundred yards from the Loch Garten road up to the viewpoint. The nest was a further two hundred yards from the 'visitor centre', which, in those early days, was nothing grander than a small garden shed. From there visitors could view the birds through powerful fixed binoculars pointing through a gap in the screen of young Scots pines towards the eyrie. The home life of these rare birds was there for anyone to see, without causing any disturbance to the ospreys themselves. It was a unique opportunity, bearing in mind that this was the only known nesting pair in Britain at that time.

Even then, in those early days, the bedrock of Operation Osprey was the team of volunteers who returned year after year to assist the wardens and help look after the visitors. I fondly remember the pleasure and anticipation of meeting up with these old friends as they returned time and time again to do their weekly stint of guard duty at Loch Garten.

Barbed wire defences at Loch Garten – it doesn't look pretty but was a necessary deterrent in the fight against egg raiders

Fifty yards further on another small hut was built, lying concealed within the trees, used by the wardens and volunteers both day and night, to maintain a guard on the nest. And, as the years went by, while the huts got bigger and more were built, the set-up remained pretty basic. At a time when optical equipment was expensive and beyond the means of most of us, we got our great views of the ospreys through big binoculars, wartime equipment salvaged from navy ships. It was rudimentary, of course, compared with today, and most of the visitors might not have known an osprey from an ostrich – but once they were in the hide and had been shown the birds, questions would flow and imaginations were captured. Many of those early visitors became eager to be personally involved in the conservation of ospreys and other wildlife. It was a superb, and the most practical, way of encouraging people to care for nature.

My first day's work as osprey warden

2nd April 1960. Up early to find myself in what looks like a terrific area. George and I first of all went to the 1959 eyrie, still in reasonable condition. I saw a pair of capercaillie by the hide (my first). Then we went to Nethybridge, met Hamish, and into Grantown to borrow ladder. Later, we cleared up the eyrie tree and laid down duckboards. After lunch, we erected the forward hide and started to erect the warning system on the tree and collected some Dannaert barbed wire to protect base of tree. This took us all afternoon.

In addition to the barbed wire at the base of the trees, electronic warning systems were also affixed to the nest trees, 1962

During the night the public observation post was closed and staff used the forward hide to guard the nest. Binoculars and various warning devices were connected from the eyrie tree to the hide. The wardens lived in a camp among the pinewoods near the croft of Inchdryne, which later, by chance, was to become my home. The base of the tree was swathed in barbed wire, usually coiled Dannaert wire 'borrowed' from the military. It did not look pretty but it served as a deterrent and, on several occasions, as a training exercise for soldiers sent to install it. In addition to the wire, electric warning devices were fixed on the tree to detect anyone trying to climb it, who would break a circuit and trigger alarm bells in the forward hide.

Each year, well before the ospreys returned, the wardens would get the site ready, open the hides and prepare the camp. Life at Loch Garten would then settle into its usual pattern and, for most years, young birds would be reared to fly the nest. But, still, there were problems in store.

In 1964 vandals tried to cut through the trunk, which had to be secured with steel stays to keep it from falling. But in 1971 came real disaster. The nest was raided during the night and, once again, the eggs were stolen. It was a terrible blow for all concerned. The egg thieves were caught as they left the forest but the eggs were never recovered. The following winter the decision was taken to remove the big side branches of the tree and to cut down the adjacent old dead tree to help prevent future robberies. The osprey tree looked terrible; stripped of its branches below the last fifteen feet, about ten feet of the upper trunk was covered with black anti-climb paint and wrapped in even more barbed wire. The needles though stayed green and, thankfully, the ospreys returned in 1972 to breed. In 1976 the osprey reserve and the two nearby lochs were purchased by the RSPB.

Scottish Daily Express

Arrow shows the ospreys' nest

Fury over osprey egg raid

By ALEX MAIN

THREE irreplaceable eggs are missing after a daring raid on the "impregnable" osprey eyrie at Loch Garten in Inverness-shire.

And last night, as police charged two men under the Protection of Birds Act, an expert said:

Egg theft at Loch Garten, reported in the Scottish Daily Express *on 18 May 1971*

Ringing Loch Garten ospreys

15th July 1973. Wind light, sunny with hazy mist. Quite warm. Across to Loch Garten by 9 a.m. – ITN and Paris Match already there – and to the tree – climbed tree and took chicks down to ground to be photographed and ringed – ring numbers: M9942 and M9943. Adults rather shy and not aggressive. Chicks back into nest after only nine minutes – back to the hide with the film crew and photographers. Female quickly to nest.

Boat of Garten – the osprey village!

In February 1986 vandals cut down nests in spite of the anti-theft defences. Once again, though, Loch Garten's staff turned tragedy into a happier outcome by re-building an artificial nest raised high off the ground that continues to be used to this day

OSPREY BACK AT LOCH GARTEN

R.S.P.B. WATCHERS were delighted to see the return of an osprey to the eyrie at Loch Garten at the weekend after last year's disaster when the eyrie was raided and robbed.

The skies have been scanned this Easter, as the ospreys usually reach the Highlands in the first week in April.

Mr Roy Dennis, Aviemore, Highland officer for the R.S.P.B., in charge of the eyrie, said: "We are certainly relieved to see this osprey and hope it will soon be joined by its mate, when they will start nest building."

Mr Dennis said there had been reports of an osprey having been sighted further down the valley, but added that it might be the same bird. Since the return of the osprey to Loch Garten in 1959, after being extinct in Scotland for about 50 years, 23 young have been reared at the eyrie, he said.

Security at the eyrie has been stepped up and volunteer wardens from many parts of Britain will shortly mount a 24-hour guard on the area.

The nest is on top of a 40ft pine tree from which the lower branches have been removed. The Army has laid a barrier at the base of the tree and electronic devices have also been installed to alert the guards if an attempt is made to raid the eyrie.

'Ospreys back' amidst heightened security after the previous year's disaster

Visitors viewing ospreys, 1960s

Loch Garten visitors' cars in June 1972 … the centre was in need of a car park

The severe drought of 1976 killed the old osprey tree at Loch Garten and the top became unsafe. Eventually, in 1980, the ospreys moved house and built a new eyrie in the nearby, live pine and remained there for a while. Sadly, vindictive egg thieves had not finished with them. In the winter of 1986, the osprey nest itself was cut down and destroyed while, in 1990, the

The millionth visitor to Loch Garten Osprey Centre, Mrs Jean Brearley, arrives in 1983 amidst much excitement and media interest

hide was burnt to the ground. After the destruction of the nest, Loch Garten staff re-built the tree using wooden struts, allowing an artificial nest to be raised high off the ground. That spring the ospreys readily adopted the man-made nest and it became a real success story, remaining in use ever since.

After the nest robbery visitor numbers rose to 60,000 a year. In the 1960s Loch Garten became established as one of the first large visitor attractions in the Scottish Highlands and

The new Abernethy Forest RSPB visitor centre – designed to be in harmony with its surroundings

Inside the Loch Garten RSPB visitor centre

became famous all over the world. Publicity went into overdrive and the local village of Boat of Garten became 'the osprey village'; the shops, teashops, garages and hotels, as well as the railway station, all seeing the benefit. The village sign soon boasted an osprey and the bird had become a brand.

The numbers of visitors to Loch Garten soon created a parking problem in this previously quiet corner of the Highlands and a new car park was built. In 1983, while I was still working for the RSPB, the millionth person to visit the Loch Garten ospreys found themselves accosted by photographers and excited RSPB staff. Mrs Jean Brearley of St Neots, Cambridgeshire, was presented with a Caithness Glass bowl engraved with an osprey by Tom Weir, the famous Scottish writer and mountaineer.

In the 1990s a closed-circuit television camera was fitted to the top of the dead tree, relaying very close-up images of the ospreys to screens in the visitor centre. Finally, in 2001, a top-of-the-range visitor centre was built looking out towards the famous nest. But, while all this may have been a long way from those early garden sheds, one thing remained constant: the volunteers still came, returning as faithfully as the ospreys themselves, to offer what assistance they could during the breeding season.

And these birds are far more than a tourist spectacle. Their every movement over the years has been recorded in the RSPB log books, providing a rich source of information which is becoming, as the years go by, an historical record in its own right. The data has been analysed several times, most recently by Andy Simkins in 1999. The nesting site to the north of the loch has now been continuously occupied for 49 years and eggs have been laid in 47 of those seasons. Throughout that time there have been at least six different breeding males, with one bird known to have returned for twelve consecutive seasons. In the early period it is possible that the same individual male bred for 17 years, but we can't be sure of that. There have been at least ten different females. One returned for nine consecutive years but four lasted for only two years or less. The latest pair, Henry and EJ (named from their colour rings), have become famous enough to have their own website.

This old established nest is often visited by intruder ospreys, which can lead to conflict: on two occasions the resident pair have been driven out by newcomers in April. The nest has twice been robbed and eggs twice blown out by high winds. Two breeding adults have died in different years while, in 1997, the birds were subjected to tawny owl attacks during the night. Clutch size has varied from the standard three eggs laid during 33 of the breeding years to two laid in four of the nesting seasons. Exceptionally, either one or four eggs have been laid, both recorded on two separate occasions in different breeding years, and, in two years, the late-arriving male has scraped out the female's early eggs (see table opposite).

The saga of Henry and EJ

Some ospreys become famous in their own right and probably none more so than the present pair at Loch Garten. Their trials and tribulations over the last few seasons have been chronicled and compiled graphically by the osprey wardens at Loch Garten on the RSPB Loch Garten website. A live webcam and the RSPB website inform visitors of the ospreys' daily lives, and emailed questions and comments have arrived from all over the world. Their saga has made the national newspapers and the evening TV News, so these birds are known throughout the UK and beyond. Nowadays, often a visitor's first question on arriving at the RSPB Visitor Centre is: 'How are Henry and EJ?'

As mentioned, their names come from the inscribed colour rings placed on their legs when as young in their nests. I ringed Henry (ochre–black HV) as a chick on the Black Isle near Inverness on 9 July 1998 (one of two chicks) while EJ (white–black EJ) was ringed by Keith Brockie in Perthshire in July 1997. I am grateful to the wardens for information, and especially to Helen Wycherley (a volunteer and a Henry–EJ fan), who helped make my account more immediate.

The individuals

Henry was hatched at Nest D03 on the Black Isle. This nest has an interesting history. It was an artificial nest which I built in a low Scots pine in a forest bog in the late 1970s. I had nearly given up hope of it ever being used, when a pair of young ospreys took it over in 1986; they built up the nest but did not lay eggs. Next spring they returned and laid three eggs and reared two young. Over the next ten years this pair was successful every year and reared another 17 young. In 1998 they had two more; on 9 July Colin Crooke climbed the tree and lowered the chicks to the ground in a bag. We ringed them. Both were males. I recorded that one was ringed with a BTO ring 1348956, colour ring ochre–black HV, placed on the left leg; wing measurement 370 millimetres, tail 160 millimetres and weight 1465 grams – I thought it might be a female! I was wrong, it was Henry – we had no further sightings until he arrived at Loch Garten five years later. Although he may have been seen briefly there as an intruder on 6 and 12 August 2001.

EJ was ringed as chick in a nest near Bridge of Cally in north-east Perthshire in July 1997; Keith Brockie, the famous wildlife artist, placed a colour ring white–black EJ on the right leg. She was first found breeding at Nest A02 in Rothiemurchus forest, not far from Loch Garten, with an old male osprey – 'orange–black VS'. She ousted his female in mid April and was incubating three eggs on 26 April, but by 9 May the nest was empty. Fearing egg thieves, I checked the nest and in my view the eggs had been predated by pine martens that had built their den at the bottom of the nest.

History of the Loch Garten nest site

The following table presents valuable data from the Loch Garten nest site (derived from RSPB sources and from the author's own records)

Year	Nest	Date male arrived	Date female arrived	Identity of male	Identity of female	Number of eggs	Young fledged	Running totals	Comments
1954	a	yes	Yes			yes	2	2	
1955	a	yes	Yes			yes	0	2	robbed
1956									
1957									
1958	a	1 May	3 May	new?	new?	2	0	2	robbed (03 June)
1959	b	18 April	22 April	same?	same?	3	3	5	
1960	b	8 April	18 April	same?	same?	yes	2	7	
1961	b	8 April	9 April	same?	same?	3	3	10	
1962	b	13 April	11 April	same?	same?	3	1	11	
1963	b	14 April	13 April	same?	same?	yes	0	11	storm damaged
1964	b	10 April	8 April	same?	same?	3	3	14	tree partly cut down
1965	b	2 April	6 April	same?	new?	2	1	15	
1966	b	4 April	2 April	same?	same?	yes	0	15	storm damaged
1967	b	6 April	8 April	same?	same?	3	3	18	
1968	b	5 April	30 March	same?	same?	3	2	20	
1969	b	2 April	3 April	same?	same?	yes	2	22	
1970	b	3 April	5 April	same?	same?	3	3	25	
1971	b	9 April	8 April	same?	same?	3	0	25	robbed (17 May)
1972	b	31 March	8 April	same?	same?	3	2	27	1 broken egg
1973	b	27 March	26 April	same?	new	3	2	29	
1974	b	9 April	13 April	same?	same	3	3	32	
1975	b	18 April	10 April	new	same	1	0	32	eggs failed to hatch
1976	b	10 April	13 April	same	same	2	2	34	
1977	b	12 April	15 April	same	same	3	2	36	
1978	b	17 April	16 April	same	same	3	3	39	
1979	b	10 April	12 April	same	same	3	3	42	
1980	c	19 April	19 April	new	new	3	2	44	old pair ousted
1981	c	29 March	3 April	same	same	3	2	46	
1982	c	7 April	28 March	same	same	3	2	48	
1983	c	10 April	4 April	same	same	3	2	50	
1984	c	1 April	4 April	same	same	3	2	52	
1985	c	2 April	31 March	same	same	3	0	52	male died
1986	c(a)	3 April	20 April	new	same	2	0	52	tree vandalised
1987	c(a)	22 April	22 April	same	new	1+	0	52	
1988	c(a)	12 April	12 April	same	same	3	3	55	
1989	c(a)	9 April	10 April	same	new	3	1	56	
1990	c(a)	5 May	5 May	new	new	3	2	58	2 young later died
1991	c(a)	13 April	27 April	same	same	3	2	60	
1992	c(a)	9 April	11 April	same	new	1	1	61	
1993	c(a)	30 March	31 March	same	same	3	2	63	female died (July)
1994	c(a)	1 April	3 April	same	new	3	1	64	
1995	c(a)	5 April	10 April	same	same	3	3	67	
1996	c(a)	3 April	4 April	same	same	3	3	70	
1997	c(a)	1 April	1 April	same	same	3	1	71	tawny owl attacks
1998	c(a)	31 March	3 April	same	same	3	3	74	
1999	c(a)	27 March	27 March	same	same	4	3	77	
2000	c(a)	27 March	27 March	same	same	3	2	79	
2001	c(a)	1 April	1 April	same	same	3	3	82	
2002	c(a)	11 April	28 March	various	same	0	0	82	non-breeding
2003	c(a)	late	late	new	new	0	0	82	non-breeding
2004	c(a)	3 April	26 March	same	same	3	3	85	
2005	c(a)	28 April	29 March	same	same	4	0	85	late returning & intruders
2006	c(a)	10 April	26 March	same	same	3	3	88	male VS early
2007	c(a)	24 April	4 April	same	same	4 + 3	2	90	male VS early

In 2002 EJ briefly appeared at Loch Garten on 4 April – she even arranged a few sticks on the nest, before being chased off by the old female. That same day I identified her at her previous year's nest in Rothiemurchus, mated with 'orange–black VS', she laid three eggs and they went on to rear one young, which I ringed on 6 July.

Orange VS (orange–black VS) – referred to as the villain at Loch Garten: I caught and ringed this bird at Rothiemurchus fish farm on 16 August 1994, and considered him to be a two-year-old. He weighed 1530 grams; the following year on 5 May I re-trapped him at the fish farm (his weight now 1480 grams). We fitted a tiny tail mounted radio so that Jason Godfrey could research his fishing journeys. We found that he had a new nest in Rothiemurchus Forest, where he and his mate reared one young; in 1996 his mate laid four eggs but they were robbed. He moved nearby to another nest, A02, which he took over in 1997 and has bred there ever since, and with EJ in 2001–2003. He's had chequered success – he has reared eight young, his eggs were taken twice by human egg thieves and twice predated by pine martens, while in 2005 his mate died and his two starving young were translocated to Rutland Water under licence.

Orange VS (this page) and neighbouring male Green JS (opposite) fishing at Rothiemurchus (photos courtesy of Bill and Kevin Cuthbert)

The 2003 season at Loch Garten

Olive, the old Loch Garten female osprey, arrived on 6 April 2003 for her tenth successive year. She had lost her old mate Ollie in the winter of 2001/2, after his twelfth summer breeding at Loch Garten. On 11 April a new male arrived at her nest, ringed 'ochre HV', yes Henry V, who later found fame and was hence named in banner headlines by the local newspaper: 'Henry Vth returns to claim his throne!' after an article by the RSPB warden, Richard Thaxton, who had used the soliloquy from Shakespeare's *Henry V* in the Osprey Diary. His name was very soon shortened to Henry. He was five years old and regularly delivered fish to Olive, who accepted him as her new mate. Mating occurred daily and Olive was expected to lay eggs around 25 April.

On 16 April one of the four vanquished males from last year ('black 6T') attempted to take the territory during the late afternoon and evening. After a lengthy battle Henry successfully defended his nest. On 17 April Henry was challenged yet again, this time by an unringed male. Undoubtedly worn out from the previous day's battle, Henry lost and a new territory holder had taken over.

This new male was unsuccessful at mating and in bringing fish for the female, and the breeding attempt failed during April and May. As a result, Olive left on 2 May. On 11 May Henry returned and attempted to re-take the territory from the unringed male. His energy apparently restored, he

succeeded. Unfortunately, Olive failed to return and find Henry in residence at Loch Garten, so in the meantime Henry courted two unringed females which were nick-named Faith and Hope, but these females only stayed a maximum of two days and Henry was left alone.

On 16 June a six-year-old female, 'white EJ', returned to the Loch Garten nest: she had been there from 26–29 March 2003, before Olive's arrival, and her path had nearly crossed with Henry's. EJ had bred again with Orange VS at the Rothiemurchus nest, where she reared one young last year. The eyrie had been destroyed in winter storms and I re-built it on 31 March. She laid three eggs and was incubating on 29 April, but she was attacked between 21 and 28 May by a big female with a green ring and the eggs were broken. Ousted, she moved in the end to Loch Garten. EJ and Henry stayed together for the rest of the summer and Henry was last seen on 7 September. Unfortunately, on 3 August 2003, his plastic ochre ring cracked and fell off, leaving Henry with only a metal ring (his orange eyes and eye-stripe are now the means by which he is identified).

The 2004 season

On 26 March EJ returned from migration to the Loch Garten nest. Three days later the first male arrived at the site. She immediately allowed the male to mate and it proved to be her former mate from her Rothiemurchus nest ('orange–black VS').

Henry arrived back on 3 April and took just ten minutes to fight off his rival, who returned to his old nest and mated with 'green 7B' (Ollie's grand-daughter). From 4 April EJ was attacked every morning for nearly two weeks by an unringed female osprey. She spent up to seven hours a day away from the nest attempting to see the other bird off her territory. The attacks were repeated day after day and, for the third year running, it looked unlikely that any eggs would be laid.

On 17 April the battle between the two females culminated in three separate attacks during the morning, more ferocious than anyone had ever seen before. At 3.15 p.m. EJ removed a broken egg from the nest and finally defeated her rival. The egg must have been laid that morning, possibly by the intruder. On 20 April EJ laid another egg and a later egg count revealed that she was incubating three eggs. Once they had settled down life returned to normal with Henry fishing for himself and his mate while EJ incubated her eggs. By then visitors had started to discuss appropriate names for the female, and readers of the RSPB blog were encouraged to suggest one. As the website proclaims:

> An update on the name for our female ringed white 'EJ':
>
> It is now 'neck and neck' between Eliza and EJ.
>
> 'I am 14 years old and I think that her name should remain as "white EJ", I think that it makes her sound like a warrior and with the recent egg smashing incident I think this is very fitting.' – Hannah from Shropshire
>
> 'I vote for White EJ. It sounds like an old Indian name to me, while Eliza sounds like an English schoolgirl.' – Dicky from Holland
>
> Sue seems to sum up what many of you have said in your e-mails: 'My vote is to keep the name EJ. To be honest, it doesn't matter what the final outcome of the

> vote is, to me she will always be EJ. It is funny, whereas the name Henry suits "our Henry" and kind of rolls off the tongue easily, none of the other names suggested for EJ have that ease or familiarity. Henry and EJ – long may they reign.'

So, by overwhelming public demand, EJ it was.

The first chick hatched on 26 May and the last on 31 May. I ringed the three young on 5 July with yellow–black colour rings 27, 28 and 29. Sadly, 28 was rather a runt, but he survived. The runt became quite a character in his own right as he fought for survival. He often only got scraps of food and put up a brave fight against the constant bullying from his older siblings. They used to attack him, pull the feathers out of the back of his head until he bled and became bald. He was affectionately called 'Baldrick' by the volunteers, as he was bald, a little runty, probably smelt a bit of fish, and his 'cunning plan' was to survive! The oldest chick flew on 17 July, the middle one on 22 July and, lastly, Baldrick flew on 3 August, 18 days after the eldest chick had fledged. EJ herself departed on 10 August and the chicks between 10–12 September, when Henry, too, left Loch Garten on migration.

The 2005 season

In 2005 EJ arrived back on 29 March followed by different males, with no sign of Henry, and many of the volunteers were worried that he might have perished in the storms over Spain during migration. In the meantime a new male 'red–white 8T' from a nearby nest, became dominant and EJ started laying eggs at the end of April but not incubating. She would often sit in the adjacent tree and look miserable as she waited to be fed. Unfortunately during this season, an egg fell out of her nest on 28 April. 'Red–white 8T' was a hopeless provider of fish and EJ would often call for hours whilst he ate a fish in a nearby tree. Then, on the evening of 28 April, Henry arrived back at the nest, at least three weeks late. He must have been a victim of the early April storms and bad weather over Spain. Strong winds on migration had probably taken him out to sea but he had battled his way back to land, needing weeks to recover his strength and head back to his eyrie. He had no body fat, and his chest, right under-wing and leg were covered in oil, so his journey must have been awful. As the website said: 'How relieved we were to cross him off our "Missing in Action" list.'

Henry fought for the nest and his mate EJ, and, after a great struggle, drove off his rival. He soon started scraping in the nest, and in his eagerness he kicked the last of EJ's four eggs out. His arrival was too late in the season for the pair to be successful in their breeding attempt but they stayed together for the rest of the season defending their nest from intruders and strengthening their bond by mating and Henry providing fish. When they finally migrated, EJ looked beautiful, and Henry had regained all of his strength and body fat. The oil had washed off during the course of the season, so Henry also looked healthy just prior to migration.

The 2006 season

Early bird EJ arrived back on the nest on 26 March with a very lively rainbow trout in her talons, the same date as in 2004, and three days earlier than in 2005. Within ten minutes of her return in the early afternoon, a male was displaying above her, and it quickly proved to be 'orange–black VS', her old mate from 2001–2003 and the 'resident' male from the Rothiemurchus nest near Aviemore. He had arrived there on 24 March.

They acted as a pair for the next 16 days, building the nest, mating, and the male bringing in fish. Then, on 10 April, Henry returned at 2.30 p.m., landing beside the pair on the nest. He was soon displaying 1000 feet above the visitor centre, calling 'pee-pee-pee' and diving and up down in a spectacular roller coaster flight. Twenty minutes later he dived to the nest and, after three hours of fighting, VS was sent packing, going to settle with his female at his old nest. The first egg was laid on 14 April and incubation of three eggs was eventually underway.

The rest of the season was encouragingly normal after the ups and downs of recent years. The eggs hatched on time and the three chicks were ringed on 29 June with yellow leg rings with black letters and numbers 8U, 8V and 8W. By late August a successful season was drawing to a close, EJ left during the second week of August followed within 48 hours by the eldest two chicks. The youngest chick, 8W, was the last to leave, and Henry waited for about three days just to make sure, and then he too left on his migration.

The 2007 season

In the evening of 4 April, EJ landed gracefully on the nest splendidly ignoring the pair of ravens that were hounding her arrival. She was later than usual but the weather in Spain had again been bad. Due to the lateness in the day there were only two visitors in the Centre, but the Osprey team and volunteers were delighted to see that she was in magnificent condition considering her migration. Next day just after dawn, Orange VS displayed for just under a minute, and was then delivering moss and sticks in earnest to the nest for the rest of the morning. He had been at his normal nest since 30 March, and once again he travelled between the two nests.

EJ laid her first egg on 18 April despite her tenuous relationship with VS and another un-ringed male. On 21 April she had two eggs and her incubation was lonely and often without supplies of trout. Next day, as the blog reports, 'The storm arrived at half two with a high-pitched peeping call above the Centre. The visitors rushed out to see a male displaying to EJ. Moments later, the male landed on the nest. All eyes were on the close-up monitor, the orange eyes and lack of coloured leg ring gave the game away. Henry was back! Staff and visitors were ecstatic'.

Just as in 2005, he started to scrape out the nest cup, and kicked two eggs out of the nest. There was shocked silence in the Centre and hands over mouths in disbelief. Henry started bringing in fish for EJ and on 24 April two more eggs were laid. Short-lived joy for the wardens because Henry kicked them out of the nest in no time.

The fish kept coming and the pair settled down; would she lay more eggs or not? We have only had a few cases of relays (where ospreys have laid a second clutch) in Scotland, but, on 11 May, a replacement clutch was started. EJ laid three eggs as her second clutch. By now the Loch Garten pair were even more famous – their exploits reported on TV, radio and in the press, and on the RSPB website.

On 14 June the first chick was seen at dawn – a truly momentous event. Unfortunately, next day, Henry arrived back at the nest from an attempted fishing trip, the result of which ended in him entangled in fishing line. It was initially reported that as Henry desperately tried to free himself from the nylon fishing line he stood on the chick and killed it. However, the footage was watched over and over from that second morning and it is now thought that the chick died in the nest prior to Henry's return. It was so sad; eventually EJ removed the chick from the nest, and dropped the body away from it.

The second egg hatched on 17 June but died next day, probably a result of the cold wet summer weather and the small size of the replacement eggs. Again EJ was seen to remove the dead chick from the nest. Finally, on 20 June, EJ returned to the nest and started to peck at the last remaining egg – it appears that there was a

Rare photos of Henry – a difficult feat due to the proximity of the nest – taken with a small digital camera from a distance of 250 yards from the nest, and using the large binoculars in the forward hide as a 'digiscope' (photos courtesy of Helen Wycherley)

Henry at dusk, Loch Garten (photo courtesy of Helen Wycherley)

hole in the shell. The chick was dead in the shell. The pair stayed together, with EJ heading off south in early August and Henry departing later in the month.

Interestingly, this year an intruder adult male bird was seen to land in the adjacent tree to the nest and perch there for about ten minutes. Neither Henry or EJ seemed that bothered, and only occasionally alarm-called. The intruder did not even get scared off by the camera movement which lead the Centre to think it could have been one of the young from 2004 returning to Scotland for the first time looking for territory and a mate (nearly all intruders fly off when the camera moves). A leg ring was not identified as he was perched on one leg with the other hidden, and when he finally tried to land on the nest it was only then that EJ saw him off, but even then it was very gently done. Hopefully next year we will get reports that one of the chicks from 2004 has survived and started breeding.

Henry and EJ have now become very famous, and their saga at the RSPB's Loch Garten reserve has touched the hearts of people not only in Scotland and the UK, but also from all over the world via the webcams, diary and blog. Everyone is now eagerly waiting through the dark cold winter months of Scotland to see if they will return in 2008, and for how many more years, or whether we will have to welcome a new pair to the nest at Loch Garten. We can be certain that the saga of this nest will keep people enthralled for many years to come.

Postscript
In 2008 EJ returned but Henry did not, so Orange VS took over the nest.

EJ and Orange VS at Loch Garten, 2008 – Orange VS is on the right-hand side. A rare and difficult shot due to the distance of the nest from the photographer (photo courtesy of David Shaw, www.davidshawwildlife.co.uk)

The second pair in Badenoch and Strathspey

In 1963 birdwatchers discovered a second pair of ospreys nesting in Inshriach Forest, on the western side of Aviemore. Its existence very quickly became public knowledge. That year the pair at Loch Garten had lost their eggs in a gale and so Operation Osprey was moved to Rothiemurchus, from where we could guard the new birds instead. Colonel Iain Grant very kindly came to the aid of the RSPB and gave the staff the use of Milton Cottage, on the road to Loch an Eilein, as their base. From here a protection scheme was put in place, an exciting extension to our work at Loch Garten. A small wooden hide was erected at the edge of the pine plantation on Forestry Commission land. The nest was located less than 200 yards away in a stark dead pine tree standing on a small hillock of cleared ground.

Two wardens would go on duty in the evening and remain for 24 hours, each taking turns to watch over the nest, sleep, cook and keep an eye open for people who might cause a disturbance. The few birdwatchers who found the site were encouraged to use the hide, but no attempt was made to open this one up to the public, the RSPB being certain that the Loch Garten nest would once again be used in 1964. We obtained exceptionally good views of the nesting birds throughout the incubation period but, sadly, the two eggs failed to hatch. The male osprey had a ring on his right leg and with the aid of a telescope I was able to read some of the numbers, identifying it as Swedish in origin. The male was then identified as a bird that was ringed as a chick in 1960 in a nest near Stockholm by Dr Sten Osterlof, the director of the Swedish ringing scheme at Stockholm Museum, himself an expert on ospreys. It was a great thrill for me, some 15 years later, to be shown that same nest by Sten. It was our first confirmed case of a Scandinavian osprey nesting within the slowly growing Scottish population. In 1964 the pair moved further west and then, in 1967, they nested in Rothiemurchus. Sadly, though, throughout all of those seven years, not one of their eggs hatched.

Reading first Swedish ring on a Scottish breeder

Friday, 24th May 1963. I was up at 5 a.m. Frosty start; no wind; beautiful blue sunny day. Along to hide at new nest with the big telescope. Plenty of rabbits, a greenshank calling over the marsh and a female capercaillie by the hide. From the hide read the metal ring on the left leg of the male osprey. Last two numbers were 08 and it was an aluminium clip ring, probably Swedish. Could not read any more of it as definition poor.

Loch of the Lowes

In March 1969 Loch of the Lowes became the first reserve ever to be bought by the Scottish Wildlife Trust. It was acquired primarily for the importance of its aquatic habitat: it is a mesotrophic loch supporting a rich aquatic flora, including slender naiad, an internationally protected species. Ospreys appeared unexpectedly just a month after the purchase, in April 1969, and first laid eggs the following year. At that time there were only four other known breeding pairs in Scotland. Just weeks later, though, a May gale destroyed the newly constructed nest, along with its clutch of two eggs. A man-made eyrie built ready for the birds' return the following year was accepted by the ospreys and the first two chicks were raised at Loch of the Lowes in 1971.

The Scottish Wildlife Trust decided from the outset, just as with Loch Garten, to protect the nest, not by trying to keep it quiet, but by inviting the public to the site. To do this they needed to employ staff, construct a hide (and later a visitor centre) and to conduct 24-hour surveillance of the nest, using local volunteers. This small 'army' still operates today, helped greatly by modern surveillance equipment. Considering the continuing public excitement about ospreys and the Loch of the Lowes' extremely scenic location on the Highland–Lowland boundary, it is not surprising that it remains a popular visitor destination with around 30,000 visitors per season. Well over a million visitors have seen the ospreys produce 63 healthy chicks between 1972 and 2007.

Loch of the Lowes

Osprey at Loch of the Lowes nest (photo couresty of David Shaw, www.davidshawwildlife.co.uk)

Nesting details at Loch of the Lowes

Year	Nest	First seen	Identity male	Identity female	Number of eggs	Young fledged	Last seen	Running totals	Comments
1968									one bird only
1969	a	Mid-April			0	0		0	non-breeding pair
1970	a	04 April			2	0	0		storm damaged nest
1971	b	15 April			yes	2	12 September	2	
1972	b	06 April	new		yes	3	9 September	5	Scandinavian male
1973	c	19 April	same		yes	2	9 September	7	
1974	c	06 April	same		yes	2	2 September	9	
1975	a	06 April	new		0	0		9	non-breeding
1976	c	04 April	same		yes	0	9 July	9	2 females laid
1977	c	12 April	same		0				one adult only
1978	c	17 April	same		0		5 September		non-breeding
1979	c	12 April	same		yes	2	20 September	11	
1980	c	29 March	same		yes	3	4 September	14	
1981	c	02 April	same		yes	3	6 September	17	
1982	c	01 April	same		3	3	3 September	20	
1983	c	03 April	same		yes	0		20	2 chicks died
1984	c	20 April	same			0			only male
1985	c	04 April				0			non-breeding
1986	c	03 April				0			non-breeding
1987	d	04 April				0			non-breeding
1988		28 March				0	9 September		
1989		04 April				0			
1990	d	01 April				0			adult only
1991	d	07 April		new	3	2	11 September	22	
1992	d	01 April		same	3	2	8 September	24	
1993	d	02 April		same	3	3	13 September	27	
1994	d	31 March	new	same	yes	3	12 September	30	
1995	d	01 April	green 8	same	Yes	3	5 September	33	
1996	d	29 March	green 8	same	Yes	3	5 September	36	
1997	d	29 March	green 8	same	Yes	3	29 August	39	
1998	d	30 March	green 8	same	Yes	3	3 September	42	
1999	d	27 March	green 8	same	Yes	2	30 August	44	
2000	d	24 March	green 8	same	Yes	3	4 September	47	
2001	d	26 March	green 8	same	Yes	3	3 September	50	
2002	d	23 March	green 8	same	Yes	3	25 August	53	
2003	d	24 March	green 8	same	Yes	3	23 August	56	
2004	d	22 March	green 8	same	Yes	2	21 August	58	
2005	d	21 March	green 8	same	4	2	22 August	60	hatched 4 young
2006	d	23 March	green 8	same	3	1	29 August	61	hatched 3 young
2007	d	29 March	green 8		3	2	26 August	63	hatched 3 young

3
Recolonisation of Scotland

Osprey monitoring

By 1966 the osprey population in Scotland had risen to three pairs. From then on there was a steady increase with seven pairs rearing eleven young in 1970. RSPB staff from Loch Garten monitored all these early nests and Douglas Weir started to ring the young ospreys. I returned to work for the RSPB in the Highlands in 1971 and resumed my conservation work on ospreys, along with many other duties and projects, even involvement with the newly discovered North Sea oil industry. The following year I produced the first osprey newsletter and, as the years went by, spent early mornings, late evenings and weekends trying to keep up with the latest developments. Eventually – as a real measure of the birds' success – the newly colonised areas in Aberdeenshire, Perthshire and Stirlingshire were handed over to other osprey enthusiasts as the workload was becoming too great.

Colin Crooke climbing tree to monitor a bog pine nest

Throughout Scotland there is to this day a regular monitoring programme carried out by a group of experts and enthusiasts, who check the nests in their own areas. I now monitor the breeding pairs in the Highlands and Moray, along with my friend Colin Crooke, in work carried out as part of our conservation studies for the Highland Foundation for Wildlife. Keith Brockie, the famous bird artist, took on, along with various friends, the Perthshire monitoring and, after nearly 20 years, passed it on to Bradley Yule. Iain Macleod and Keith Duncan originally kept a check on the Aberdeenshire ospreys but this is now organised by Ian Francis of the RSPB. In south-west Scotland my old colleague, Roger Broad, now retired from the RSPB, organised the monitoring in that region. And, as the numbers grow, more people take on the exciting task of monitoring and conserving ospreys.

This annual monitoring tries to collect useful data from every occupied nest. All the nest sites have been assigned a code number, usually starting with a letter that denotes the area or colony. The sequence of occupation means that Loch Garten is designated A01 and subsequent nests in that area are prefixed A02, A03 and so on. Subsequent letters refer to other parts of Scotland in sequence of recolonisation. All the nests are also given a code number to allow us to report on the monitoring results without giving away confidential site information. Of course, each person monitoring an osprey nest knows their own particular code.

Our first visits to the nests take place in March and give us a chance to catch up with our many old friends who have ospreys nesting on their land. Firstly, we examine the nest to see if it has remained intact through the winter storms: in some cases we may find that the nest has

Nest site B01

Nest A02

been destroyed by high winds or snow, or that supporting branches have been broken. In the worst cases the tree itself may have blown down. We often re-build or stabilise the nests.

Sometimes, on this first visit, we find that one of the birds is already back and starting to rebuild the nest – our contact person on the ground may well know exactly when it returned. We record the dates of arrival whenever possible and keep a record of all our visits for each eyrie. Generally, if a pair of birds is back early, it is likely to be the same birds that nested there the previous year. On other occasions, only one bird has returned and it may have to wait a week or more

Pre-arrival early checks

28th March 1999. Morayshire.
B14. 1215 – no birds, nest ok.
B15. 1220 – nest and perch ok.
B9c. 1234 – nest ok – smaller – nil.
B4a. 1237 – nest ok but a bit slipped over – nil.
B4b. 1241 – nest ok but bit blown off – nil.
B11. 1245 – nest ok – nil.
B01. 1307 – female on nest, nest ok.
Same very heavily streaked bird.
B10. 1323 – nest a half blown out but base ok – nil.
B17. 1336 – nest OK, nil.
B02. 1403 – nest OK but thinner – nil
– buzzard landing nearby.
B3. 1415 – nest OK – nil.
B22. 1419 – nest OK nil.

Using mirror pole at nest site A01

Checking eggs at nest A11 with a mirror pole, April 2005

Meeting contacts at start of season

15th April 1971. Drove past the keeper's gibbet – the alleged 'osprey' was actually an immature gull hanging up, some big wild cats hanging up. Drove on to see keeper; nest branch blown out, both birds present. Male and female in large Douglas fir – then male with fish; later female soaring and male gliding away towards the loch. To Forres and so to keeper's house and we had tea. Nest (2) in good order – one bird above eyrie in larch. Unusual site. Later to Loch Flemington, in to see Mrs Beaton – 'their' osprey arrived Monday 12th same as other nests. We again scanned Cawdor woods but no luck.

before being joined by its mate. If the partner has died, it may take some time for a new pairing to occur.

Once a pair is in residence, we try, in as many cases as possible, to establish whether they are the same birds as in the previous year and whether they are known individuals, identifiable by colour rings or plumage characteristics. The next checks are made from mid-April to May, to see if incubation has started, and this monitoring continues into May. Where nests are easily accessible, we check the clutch size simply by using an aluminium pole with a mirror attached to the top. We also note the colour and patterning characteristics of the eggs. By this time all old nests and artificial nests have been checked for occupancy; new nests are searched for and leads from others followed up. It can take many hours to find new nests. On a few occasions I've even used a small plane or helicopter to locate nests.

Landowner and keeper at ringing time

Later, during May and June, we check to see if the eggs have hatched and, if not, try to establish why not. In late June we visit the nests to see which have chicks and we try to count the number of young so that, in early July, we know which ones we need to visit at ringing time. As many young as possible are ringed and the numbers are noted for later comparison with the number that actually fledge, late in July and August. Later visits are made to check departure dates, whenever possible. This series of visits allows us to log a comprehensive knowledge of each nest and its occupants, and enables us to produce a report on breeding success. In this way we obtain accurate annual knowledge of the productivity of ospreys in Scotland. It is difficult, nowadays, to find all the new nests and we are always interested to hear from anyone who finds a nest: it might be one that we are not yet monitoring.

As I've already said, much of our conservation work relies on the goodwill and cooperation of private landowners, factors, gamekeepers, stalkers, farmers, crofters, foresters and local folk. They are a wonderful group of interested and interesting people and I feel privileged to have known and worked with so many of them through the years.

Scottish renewal

As the years have gone by I've seen pairs establish themselves in different areas while old favourites disappeared. Ospreys are now to be found in many parts of Scotland and the unexpected thrill of seeing one of these spectacular birds fish for its prey can be enjoyed by so many more people. My personal joy is to see ospreys hunting close to the high Kessock Bridge when I'm driving across the Moray Firth between Inverness and the Black Isle, and, several times, even in the centre of Inverness, I've seen an adult bird fly over with a fish trailing below, caught in its talons. While there are still many areas waiting to be recolonised, the osprey is now secure in our country. I often wish that George Waterston, Iain Grant and the other pioneers who nurtured those first few ospreys back in the 1950s, and who are, sadly, no longer with us, could know of the success of all their early efforts. Maybe, somehow, they do.

Finding new nests

21st April 1972. Wind northeast, high overcast, lovely evening. Loch Ness. 4 p.m. male osprey arrived just west of Dores – perched on lochside tree – no rings definitely. Suddenly plunged and caught a fish, thought to be trout. Chased by common gulls – turned over Dores and made off inland. Chased in car but lost. To Inverness to do a radio programme 'North Beat' with George Pease. Then back out to Dores area – frequent stops and searches – beautiful visibility – no sign. Couple of kestrels, one buzzard, several willow warblers in song. Suddenly while scanning from high road – saw an osprey on a new eyrie about three miles away by telescope. 6.30 p.m. round by car and good view of both ospreys at eyrie in small pine tree. New nest! Home by 8 p.m.

1966 to 1980

By 1966 the population in Scotland had risen to three pairs. Really importantly, a new pair had become established in Moray but, unfortunately, summer storms claimed the clutches of both the pairs in Strathspey that year. In 1967 numbers grew again, this time to five pairs, and that first pair in Moray reared two young. In addition, two separate non-breeding pairs built nests in Moray and Strathspey.

In 1968 there were three pairs in Badenoch and Strathspey but only the Loch Garten pair was successful in rearing two young. There were three pairs in Moray and two of these reared at least four young. There was also a further colonisation, with two single birds establishing territories in different parts of Perthshire, a hugely significant step for the species. In 1969 we noted seven nesting pairs, of which five laid eggs, rearing at least seven young. The Perthshire birds both attracted a mate but did not breed.

In 1970 the population remained stable at seven pairs. Importantly, a pair at Loch of the Lowes laid eggs that failed to hatch, but the three Moray pairs were highly successful and produced a total of eight young between them. So, by 1970, a total of 44 young had been reared in Scotland and 21 of these had been ringed. Numbers were rising steadily.

By 1971 there were eight pairs, but there were also some significant new additions. There were single ospreys at two new sites in Moray while a non-breeding pair built a nest near Inverness. In 1972 we recorded a further increase to 13 pairs but there were still just three pairs in Strathspey with only the Loch Garten nest being successful. There were now six pairs in Moray but only three pairs reared young while the Inverness pair had two young. Importantly, a new pair moved a long distance away to the south-west, to build a nest near Lake of Menteith and there was also a report of a pair with three young in south-east Sutherland. The following year there were 14 pairs and 10 of them laid eggs, rearing a total of 21 offspring. Both the Perthshire pairs reared two young apiece and five out of eight pairs in Moray were also successful. The new pair in the south-west reared three young but the pair in Sutherland either failed or was non-breeding, while the nest near Inverness was destroyed by wind with only one adult being present. These gains and losses were to become a familiar part of the pattern of osprey recolonisation.

In 1974 there were 14 pairs and all laid eggs, with 21 chicks fledged. The pair in the south-west again had three young while those in Sutherland had two, and a new young pair built a nest there. But the Inverness area was empty of ospreys that year.

In 1975 there were nine pairs in Moray and their favourite fishing area was Findhorn Bay. There were still only two pairs in Strathspey and Perthshire. In a throwback to worse times, the nest near the Lake of Menteith was robbed in 1976 and, frustratingly, there were no young at Loch of the Lowes because two females laid eggs and tried to incubate in the same nest. No new areas were colonised, but a further non-breeding pair was recorded in Moray.

Good day in the field

18th July 1975. Ian and I set off to North Perthshire nest. The gamekeeper and his son came with us and rowed the boat out to the nest. Three big young. Ian lowered them down and I ringed them – M9987–89, with colour rings M, N and Z – all looked in good form (ca. six weeks) – female noisy and kept hundred yards up. Nest very big – no sign of fish – lined with sphagnum moss, flotsam from the loch and heather clods, plus one mallard feather. On the way back, I was dropped off at the point and hid under tree with my telescope – female to nest in five minutes and confirmed that colour ring is black on right leg, BTO ring left – after one hour plus managed to read part of the ring – (M9)90(3) – last number almost certainly 3 – but large branch in the way. Ian came back and we went to the top island – superb blaeberries, tufted duck nest with eggs and one big reddish adder. On second island, two half grown kestrel chicks in a goldeneye nest box and lots of beautiful twinflower Linnaeus borealis. Spotted female scoter with five half grown young as we rowed back. Great day.

In 1977 the population increased to 17 pairs, but it proved to be the worst breeding season so far because cold wet weather caused six pairs to fail during incubation. And, as if that wasn't bad enough, the first clutch of four eggs was stolen by thieves. But, at least it was a year of territorial gains, with two more nests recorded in Perthshire, a nest with one young in Aberdeenshire and a non-breeding pair in Easter Ross. These were important improvements and all the signs were encouraging.

1978 was another good year with an increase to 24 pairs rearing 20 young between them. More importantly, there were now five pairs in both Strathspey and Perthshire while the new pairs in Aberdeenshire and Easter Ross both reared two young each. During the next two years we logged another pair in Aberdeenshire, two more pairs in Strathspey, and the first pair ever recorded in Angus. The new decade started on a happy note with 26 breeding pairs of ospreys in Scotland, 20 of which laid eggs rearing a total of 41 young. This brought the running total of young reared in Scotland to a marvellous 250.

Looking for nests from the air

13th May 1978. Aerial survey. Late noon to Inverness airport – met Keith Durbridge at the flying school and out in his Cessna plane at 5 p.m. Straight along over Culbin forest looking at the Cran Loch – nil there – but female incubating at B5 – didn't get off the nest – everything OK. Checked the cartwheel at Binn's Ness – not in use and no extra building. Flew up the River Findhorn – some good looking trees on river side.
At B11 one bird incubating – good view of it – and so to nest B4 – bird incubating and not moving at close approach. B13 nest in tall Douglas fir – good new nest but no birds – then checked Castle area. Next to B2 but couldn't see anything on the eyrie. On up river – nil at Logie – found B9b nest in tall Douglas: one bird incubating on nest. On up river but couldn't find anything except one old nest in a large pine tree. Searched Glenferness area – nothing found – out over moors to Ord hill and so to Cawdor but no sightings at all on the way back to the airfield.

Early breeding success of the population

In order for their population to increase, ospreys need to produce enough young to return to breed over and above the natural losses that occur during the birds' early years in Africa and on migration. The early years of colonisation were very slow and it was not until the late 1960s that the population started to rise. Interestingly, a varying number of pairs each year take the trouble to build a nest but do not lay eggs. These non-breeding pairs are included in the population totals and they are important for the future of the species. Between 1966 and 1980 the percentage of non-breeding pairs ranged from 7% in 1974 to 50% in 1968. Of course, the fluctuations are more marked when the population is small.

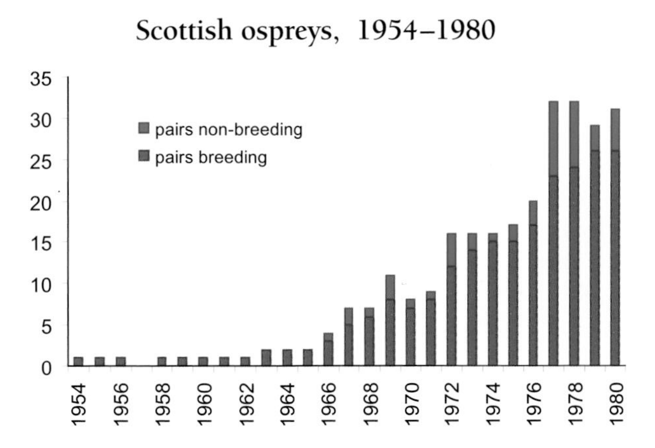

Big brood of three young in nest above a cow field

It is not just the number laying eggs that indicates population success but the total number of young produced each year that go on to fly the nest successfully. This is called *productivity* and there are three ways of measuring it. Firstly, the number of young reared from the number of pairs that are present on their nests in the spring; this includes non-breeding pairs. Secondly, the number of young reared per pair that initially laid eggs and, thirdly, the number of young in each successful brood.

In the early years, when the numbers in the colonising population are low, these productivity indicators can vary widely. For example, it only needs two pairs out of four to be unsuccessful for the indicators to crash. So if we examine just the three five-year periods between 1966 and 1980, we find that the breeding success ranged between 1.57, 1.3, and 1.06 young per pair at a nest and 1.83, 1.60 and 1.41 young per pair laying eggs. The mean or average brood size was 2.7, 2.18 and 1.95 young. These indicators were interpreted as being encouraging and they suggested that the population would continue to increase.

1981 to 1990

The 1980s saw a period of sustained growth in the osprey population in Scotland. Starting from a base of 26 pairs in 1981, the decade finished with more than double that number. In 1981 a new pair had become established in the northern part of Easter Ross, as had a pair near Inverness, who had adopted an artificial nest

Scottish ospreys, 1981–1993

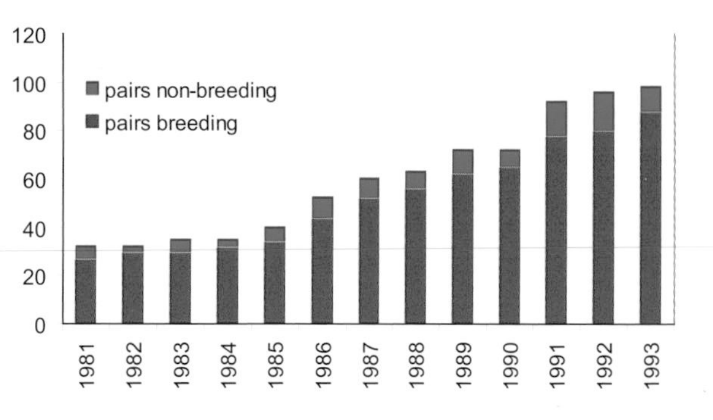

Spring field surveys

8th May 1980. Good morning at the Morayshire ospreys.

Nest B3. Female incubating: nest built up in usual live pine. Female off when we got halfway to nest, three eggs, female quite noisy. Warm brown eggs with buff background. All similar. No one had been up the tree. Female back on nest when we were half way back. No rings, perched on tree to west of nest, dark breast band and throat.

B2. Female on nest, first seen about 7th April, not very big nest. Called but shy and flew off and back, couldn't see any colour rings but female down very quickly on to eggs.

B4. Male on tall Douglas fir other side of river – with telescope could see metal ring right leg, nil on left. Climbed up carefully (to cliff top) and could see female on the nest had green ring left leg and metal right. Started to call at me and flew off nest. Three eggs, one rather pale.

B7. Pair at the nest, calling and obviously incubating, but couldn't see their legs to check for rings.

B11. The female incubating, two other ospreys flew straight over to the east, used the mirror pole and three eggs in nest. Female quite noisy and then male turned up and called but floated away to perch.

B1. Female incubating got off early as I walked to nest with mirror poles, bigger nest than ever, three eggs, very pale with dark blotches. Then male overhead with two other ospreys, female back on eggs quickly.

B10. Female incubating, male perched in tree next door; checked with mirror poles three eggs in nest, female very tame and quickly back on nest, she has BTO ring on left leg, nil on the right.

Female at nest B10

Nest B10, Moray

that I had built some years before on the Black Isle. It was gratifying to watch this pair rear two young. By 1982 there were eight pairs in Badenoch and Strathspey as well as three occupied sites in south-east Sutherland. The following year there were 30 pairs with the same distribution but, in 1984, we logged our first pair in East Moray. These birds were fishing on the estuary of the River Spey and this, in future years, was to become another favourite hunting ground. A third pair was also recorded in Easter Ross.

By 1985 the population had increased to 34 pairs, with four pairs in Aberdeenshire. The east Moray pair reared three young that year. 1986 saw a further large increase to 42 recorded pairs, 35 of which laid eggs. The new ones were two pairs in Strathspey, Easter Ross and Inverness-shire respectively and one pair in both east Moray and Perthshire. In 1987 numbers rose again to 50 pairs with another pair logged in the south-west and a fifth pair recorded in Aberdeenshire. Further increases in the next three years meant that the total reached 62 pairs in 1990. Of these, 56 pairs laid eggs and 90 young were reared, bringing the running total to 836 young ospreys known to have flown from Scottish nests since 1954. We could truly begin to believe that the birds were here to stay.

1991 to 2000

The population rose to a grand total of 72 pairs in 1991 but it was a poor breeding season. Although the first bird was at its nest on 17 March, many were late arriving due to bad weather in Spain. Further poor conditions in Scotland then compounded the problem and caused many nests to fail.

1992, though, was a milestone year with 101 youngsters reared. This was the first time a hundred young ospreys had flown the nest in a year in Britain for probably more than two centuries. There was an increase in conflict among the birds with intruders at occupied nests resulting in egg breakages, a trend that has continued. Thankfully, though, the number of egg thefts declined.

Just four years later, in 1996, another milestone was reached when the year's tally of 104 nests was the first time that there had been 100 or more breeding pairs in Britain for several centuries. It was so encouraging to see the population rising but it was becoming ever more difficult to locate all the nests. At this time the first pair bred in the Scottish Borders, south of the line between Edinburgh and Glasgow. By the end of the decade, the population had reached 147 pairs, recorded in the year 2000, with 195 young reared. The running total had now reached at least 2209 young successfully reared to fly from Scottish nests since 1954 – a fantastic way to close the old century and herald the dawn of the new.

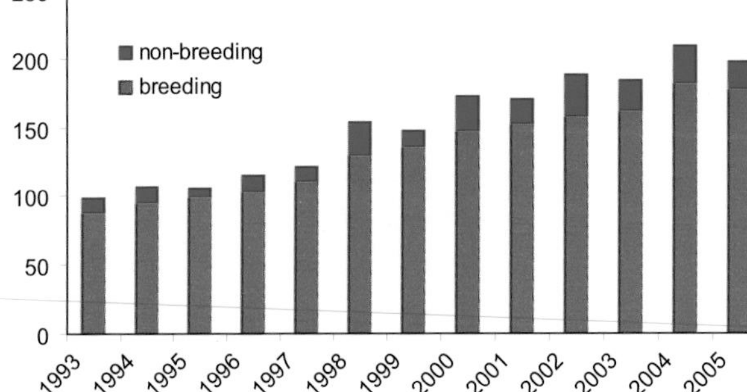

2001 to 2007

A further milestone was reached in 2001, when, for the first time, the number of young reared in a year passed the 200 mark. The total population had climbed to 153 known pairs. Four years later the population had risen again to 182 known pairs, and the first ospreys had started to nest again in Lochaber and in Galloway. In 2005 the ospreys endured a very difficult spring migration and more 'regulars' than usual failed to return. As a consequence, only 178 pairs were located but 125 of them reared a total of at least 242 young – another record.

In 2006 populations in the north and north-east showed little change, but the population in the south and south-west continued to show growth in line with normal recolonisation. Breeding success was generally high with very good numbers of young reared. Full details were not obtained for a couple of areas, but at least 184 pairs reared at least 247 young, with a total population including unchecked known nests reaching 195 pairs. In 2007 there were more losses of returning adults than usual, possibly more than double the normal 9% annual adult mortality, because the April weather in Spain and the Bay of Biscay was windier, wetter and more overcast than usual. The Scottish population probably remained the same, but there were now three pairs in Dumfries and Galloway. Breeding success was generally poor, with more failures than usual due to the wet summer weather.

And so, the process of recolonising ancestral osprey haunts continues, albeit slowly at times. For comparison, look at the speed with which the peregrine falcon bounced back from the devastation in its numbers during the 1950s and 1960s, the era of widespread pesticide use. Numbers reached rock bottom in the 1970s, but the peregrine is now breeding throughout the whole of the British Isles, in higher numbers than were ever recorded before. The peregrine, whose name means 'wanderer', has the happy ability to recolonise its former haunts much more easily than the osprey.

Overall breeding success

As numbers started to increase more rapidly from 1981, some of the annual fluctuations, seen when the population was small, started to even out. This was particularly true for the percentage of non-breeding pairs which ranged from 7% in 1982 and 1995 to a maximum of 20% in 1992. Looking at the complete data from 1954 to 2004, it can be determined that the average percentage of non-breeding pairs per year is 14% of the whole population (from a sample of 2369 pair/year).

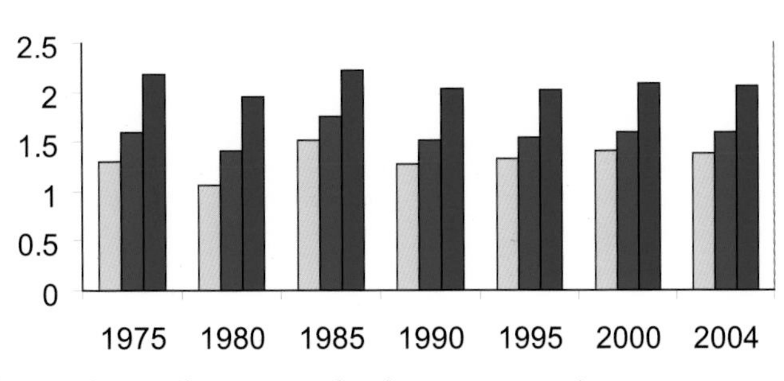

Variation in breeding success in Scotland has also been seen to even out when the population is larger and more widespread. It has also started to show a surprising similarity from year to year over the whole population. Productivity per nesting pair of ospreys varied from 1.05 young in 1991 to peaks of 1.56 in 1981 and 1985. The productivity per pair which had laid eggs varied from 1.28 young in 1991 to a peak of 1.91 in 1981. Finally, the average brood size ranged from 1.86 young in 1991 to 2.41 young in 1985.

Adding up all the figures, we find that over the 50-year period of recolonisation, the productivity per pair with a nest is 1.35 young with a range of 0.57 to 1.58. Additionally, the productivity per pair laying eggs has averaged 1.57 young with a range of 0.93 to 2.05. Finally, the average brood size has been 2.2 young per nest with a range from 1.73 to 2.75 young. These figures are likewise derived from a total number of 2369 pairs studied between 1954 and 2004.

The breeding success of the Scottish osprey has been maintained at a productivity which is excellent in comparison with that of similar countries. 1.57 young per pair laying eggs (or as some countries call it, per active nest) compares favourably with 1.46 young per pair in Finland and Sweden, 1.5 in Norway, 1.7 in Germany and 1.43 in Corsica. Early research in North America suggested that a breeding success of 0.8 young per pair would ensure a population increase. The Scottish figure is much higher than this but, even so, the population here has not increased in recent decades as much as we had expected. There are several possible reasons for this. Firstly, up to 5% or maybe even more of new breeding pairs each year may go undetected because we are not finding the nest sites. Secondly, young birds which are physically able to breed at three years of age may now be delaying breeding until they are older, probably because of competition with established pairs. Finally, the Scottish population may have a higher mortality than those that were studied in North America. This possibility is discussed later.

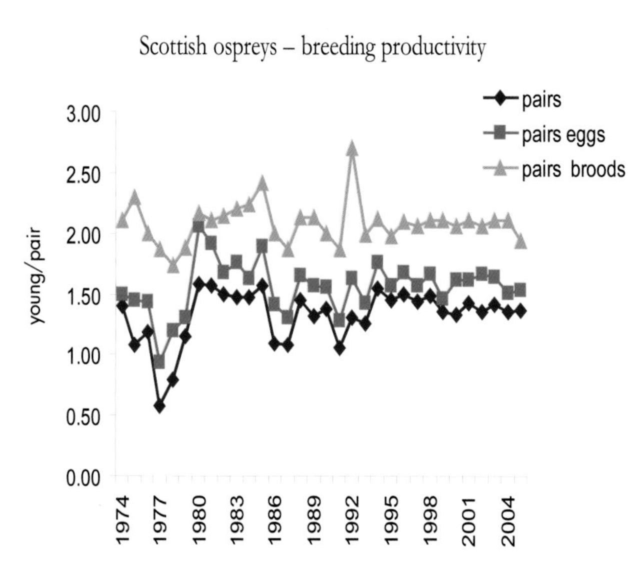

The process of recolonisation

Ospreys breed in loose colonies in many parts of the world, a process now repeated in Scotland. Adult ospreys return to the same nests year after year for as long as they survive and there is very little change of sites. The new breeders are the ones who establish themselves in virgin territories, although most prefer to breed near other ospreys. In fact, the potential breeders often try to take over occupied nests, either before the established owners have returned from Africa or by fighting with them once breeding is underway.

The first nest recorded away from the core site in Strathspey was 33 kilometres distant in Moray. That pair was then joined by others in the same general area to form a loose colony with a common feeding ground in Findhorn Bay. This pattern of recolonisation and spread of range has continued. Up until 1995, new 'colonies' were started by pairs building their nests in near-by areas, at distances between 12.5 and 67.5 km from the nearest occupied site, at an average distance of 40 kilometres. Within a few years, the pattern has usually been that of at least one other pair joining an occupied site to form the beginnings of a new colony but, should no others arrive, the process fails. The number of breeding pairs within a 'colony' increases during the early years of establishment in a new area but with time, it reaches an upper limit and may even start to decrease.

One of the most interesting and encouraging facts about the recolonisation is its failure to mirror the decreases of the nineteenth century. Although one of the last eyries used in Scotland in late Victorian times was only 13.5 kilometres from Loch Garten, the great majority of the ancient nest sites were located in the western half of Scotland. The new population spread out from Loch Garten with the central core remaining in the eastern half of the country. There is a very clear difference between the old remnant population and the new distribution; an analysis in 1995 showed that 83.3% of the ancient nests were in the western half of Scotland, while

90.5 % of the new nests were in the eastern half. This pattern agrees with our view that ospreys once nested all over the British Isles and had been persecuted to extinction from many nest sites before written records were available to document the decline. The situation in the nineteenth century reflected a remnant population heavily affected by human persecution that clung on to survival in remote areas where persecution was less severe. The expansion of range is now occurring in those parts of the country where fish are most abundant, where there is less rainfall and the underlying productivity of the land is greater. These areas are also more densely populated by humans but, nowadays, within a prevailing climate of bird conservation, ospreys can finally survive and flourish.

Young per year 1954–2005

total young

New 'colonies' distance from nearest nest (kms)

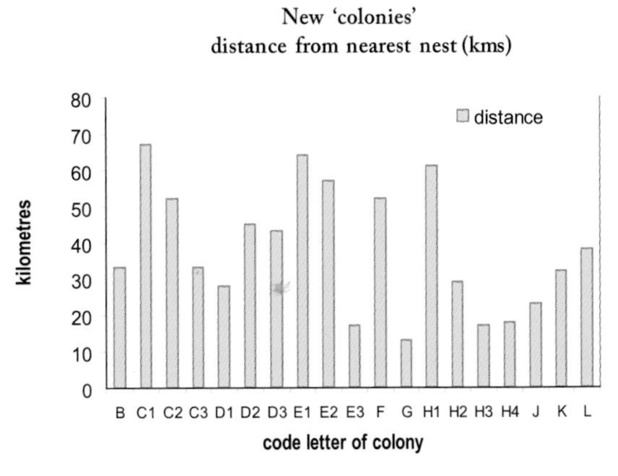

OSPREY'S NEST OF 1892, ON THE DUKE OF RICHMOND AND GORDON'S PROPERTY.

4
The history of the osprey in the British Isles

"I think he'll be to Rome
As is the osprey to the fish, who takes it
By sovereignty of nature ..."

– Aufidius to his lieutenant on Coriolanus' prospects of taking Rome by force,
in Shakespeare's *The Tragedy of Coriolanus*, Act IV, Scene VII

From ancient times to the early 1900s

The osprey has had one of the saddest histories of persecution of any creature in the British Isles. All of our native large predators – both birds and mammals – have been hunted, trapped, poisoned or shot in the course of the last millennium, and birds of prey have often been first in the line of fire. The killing accelerated during the eighteenth and nineteenth centuries with the advent of game estates, coupled with the era's prevailing antagonistic attitude towards wildlife. Anything that competed with human interests was simply not to be tolerated and relatively few people questioned the morality of the annual death toll, nor expressed concern when creatures, once common, began to disappear. The records of vermin killed on shooting estates in the Highlands in the nineteenth century are both hard to credit and hard to stomach. In Glen Garry, between the years 1837 and 1840, 98 golden eagles and 18 ospreys were killed, along with innumerable buzzards, red kites, harriers and sea eagles. This slaughter was repeated all over the country and, in addition and a terrible irony, the market for museum specimens grew as the 'vermin' themselves became rarer. No wonder that osprey nests were raided when a single egg was worth 7*s*. 6*d*. – a worthwhile sum for many a poor person on a meagre income in 1837.

I find it particularly interesting that, even by the end of the eighteenth century, the osprey had become a rare bird, yet there seems to be no doubt that it had nested throughout the whole of the British Isles at the beginning of the last millennium. In comparison with other birds of prey (with the exception, perhaps, of the white-tailed sea eagle), there seems to have been an early selective killing of fish-eating raptors, both in England and elsewhere in Western Europe. During my research into the ancestral distribution of ospreys in England, I became convinced that this early persecution in the Middle Ages was due to the birds' predilection for hunting from fish ponds. In those days, before the advent of easy transport and deep-freezes, it was difficult enough to find fish to eat on Fridays, a strict religious requirement at that time. Most big houses, castles and monasteries established their own fishponds, or, as they preferred to call them, 'stew ponds'. Ospreys, as we now know, will always go to hunt at a pond stocked with large numbers of fish – it is natural to prefer easy pickings to hunting wild prey. The keepers of the stew ponds must, I am sure, have ranked the osprey among their worst enemies and done everything possible to be rid of the birds. So, by the seventeenth century, the osprey had already been pushed back, mainly into the wild and less populated parts of the country, where fewer humans meant less conflict. There was obviously a lot of investment in stew ponds, for the official records of Pell Office reveal a payment of £66 13*s*. 4*d*. on 10 October 1618, to one Robert Wood, 'Keeper of his Majesty King James the First's cormorants, ospreys and otters' for making fish ponds and for building a house to keep the animals. Does this indicate, I wonder, tame animals for hunting, or a keeper killing vermin?

England

As indicated, the main persecution and loss of the osprey in England occurred many centuries ago. Old references suggest that the osprey was widespread in the tenth and eleventh centuries and was still common during the Middle Ages. Caius, writing in 1570, said: 'they are abundant with us on the sea coasts and in the Isle of Wight; our people call it an osprey.' In 1577 Harrison states: 'We have also ospraies, which breed with us in parks and woods, whereby the keepers of the same do reap in breeding time no small commodity: for so soon almost as the young are hatched they tie them to the butt ends of sundry trees, where the old ones finding them, do never cease to bring fish unto them, which the keepers take and eat from them.'

Ospreys were mentioned in mediaeval plays. George Peele (1552–1589) writes in his play, *The Battle of Alcazar*:

> I will provide thee of a princely osprey
> That, as he flieth over fish in pools,
> The fish shall turn their glistening bellies up,
> And though shalt take thy liberal choice of all.

Shakespeare (1607) refers to the behaviour of ospreys in his play, *Coriolanus*. No writer would have used these references unless the bird was well known to the people of the time. Carew (1602) writes that in Cornwall the 'osprey is a seafowl but not eatable' while Willoughby (1678), writing about Yorkshire, states that the bird 'preys often upon our rivers'.

By the time the first references to ospreys appear in natural history writings, the bird is in terminal decline. They bred in Cornwall at Cargloth cliff, Portwrinkle, Downderry and regularly at the Lizard in the 1700s. There were two young recorded in a nest at Langstone in 1717. In Devon they were regularly seen until the middle of the 1780s. They bred at Beer in the eighteenth century, on Lundy until 1838 (cliff nesting) and on the north coast until 1842. Ospreys last nested in the Isle of Wight in 1570. In Sussex, Knox, writing in 1849, cites the hatred reserved for ospreys by the proprietors of pike ponds. He states that he was sure the birds would nest if left unmolested and hints that they did just that before his time. There are nesting records for Whinfield Park and Ullswater up until the end of the eighteenth century. The last English record is of a pair building a nest at Monksilver, Somerset, in 1847. It testifies that one of these ospreys was shot by the keeper. One can only wonder what happened to that last, lonely survivor.

Wales

There are no written records of breeding in Wales but, just as in the rest of the British Isles, there are many areas, both inland on lakes and rivers and on the coasts and estuaries, that provide ideal habitat for ospreys. The problem again, in my view, is that the species was almost certainly exterminated before ornithological records were kept and published. During research for a Forestry Commission proposal to translocate and release ospreys in Wales, Roger Lovegrove and I examined the situation in detail. We could not find any written records of breeding ospreys but the circumstantial evidence, coupled with our knowledge of the species' ecological requirements, convinced us of its earlier presence. More interestingly, a literature search commissioned by the Forestry Commission revealed a range of historical references, particularly relating to the language.

Tym Elias found ten recorded Welsh names for the osprey, seven of which date back to the early nineteenth century. The earliest reference is from Sir Thomas Wiliems' dictionary of 1604–1607. Although there is some possible confusion with the white-tailed sea eagle, which was also originally present in Wales, there are distinctive references which undoubtedly denote the osprey. *Gwalch y mor*, *gwalch y weilgi*, *pyegeryr*, *pysgwadwalch*, *barcud y mor* and *pysgod-walch*, as well as the modern Welsh name *gwalch y pysgod*, all suggest that the bird was well known to the native people, probably right up until the eighteenth century, by which time it had become extinct as a breeder. The osprey also appears on the Coat of Arms of Swansea, arms that were granted in 1316. To me, all the evidence suggests that the bird was once a common sight in Wales.

Ireland

The situation is similar in Ireland although the references to breeding are more detailed. Gordon D'Arcy gives a range of Irish native names for the osprey – *seig*, *Iascair cairneach* (tonsured fisherman), *coirneach*, *preuchan eannan* and *iascaire ceannan*. Such a list of names shows that the bird must have been well known to the population. Interestingly, the first person to write about the Irish osprey was Giraldus Cambrensis, a Welsh monk who went to Ireland in the twelfth century. He gives a very good description of the osprey's fishing habits. Others, such as Molyneaux (1684 and 1709), Marham (1790) and Beranger (1179), provide similar accounts of ospreys, so clearly they were once breeding on the island. Molyneaux describes osprey fishing behaviour very well, while Connaught, on a tour in 1779, describes an osprey's nest with calling young, on a ruined castle on an island in Lough Key, County Roscommon. Yet in the next century, no writers mention the bird and it had become extinct. There is no doubt, though, that Ireland provides excellent breeding and fishing habitats for ospreys.

Scotland

As in other parts of the British Isles, the early history of the osprey in Scotland is lacking. I always find it strange that there are no records of the osprey nesting in Moray, since it has become in recent decades such an important county for breeding, and has always provided ideal habitat for the species. Presumably the species was exterminated in this relatively densely populated area before any written records were kept. In the nineteenth century, though, things changed, and numerous written references to the osprey appear. As persecution increased and intensified, the ornithological books of the time devoted more and more pages to the osprey.

There were probably 40–50 pairs remaining by 1850 and some chose dramatic nest sites, like the abandoned castle at Loch an Eilein in Strathspey, and the gaunt ruined towers of the castle at Loch Assynt. The ospreys could hardly have chosen a more remote and beautiful place than this ancient ruin, surrounded by the mountains of Sutherland. Whenever I visit, I love to

Loch an Eilein, Strathspey, June 1997

Ardvreck Castle, Loch Assynt

think of the time when the ospreys nested there, but, nowadays, the traffic rumbles by unheedingly on the public road which passes close by and no-one would know they had ever been there. Scottish ospreys did not confine themselves to trees (in fact, they still show great enterprise, as will later be discussed) and were used to nesting on old castles such as these, rock pinnacles and even on low rocky islands such as the osprey rock on Loch Mor Ceann na Saille. Comparison of lithographs in the natural history books by J.A. Harvie-Brown with the landscape of present day Sutherland and Wester Ross shows unchanged rocks denuded of their osprey eyries. I wonder if the bird will ever regain its former haunts there?

During the 1800s the history of the last ospreys in Scotland was becoming well documented and it is possible to trace the location of the nests at that time. Plotting nineteenth century locations on a map of Scotland reveals four old sites in the south-west, in Dumfriesshire and southern Ayrshire. A further five nests were located around Loch Lomond and in south-west Perthshire, while a further grouping of nests extending from Loch Tay and Rannoch Moor towards Loch Laggan. An isolated nest was near Dunkeld, and then seven sites in the valley of the River Spey, including the famous nest at Loch an Eilein, described above. Eight nests were in the general area of Loch Arkaig and Loch Loyne; one in Easter Ross and five in Wester Ross; one in Lewis and an important grouping of eleven nests in north-west Sutherland.

Osprey rock on Loch Mor

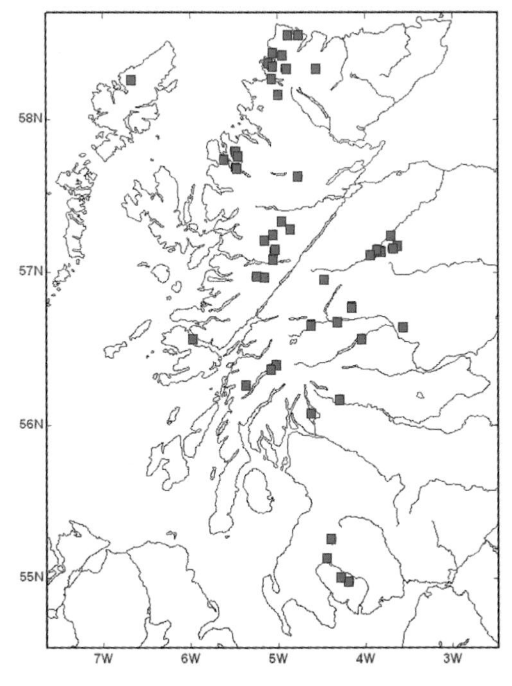

Nineteenth century osprey nests in Scotland

The distribution of the birds at that time was, as stated previously, very much in the western half of Scotland. Eighty-three per cent of historical nests were west of a line between Dumfries and Tongue, but by the 1990s, 90% of the new population was nesting east of it, in the more fertile half of Scotland. This shift in the distribution of the birds is nothing more than an indication of the pattern of ancient persecution. Ospreys had been eliminated from the more populated areas, with their rich farmland, fisheries and large estates, while in the remoter regions, where competition with man was much less severe, the removal of the last osprey pairs was due to the advent of travelling naturalists collecting eggs and specimens for taxidermy. These men made special journeys to kill the final few remaining birds and to take the last of their eggs to add to their private museums and collections. At that time an osprey egg could be sold for 7*s*. 6*d*. to a collector, a lot of money in those days. Thinking of these raiders – on the one hand so fascinated by the osprey, on the other so blind to its future – I wonder at the sheer energy that went into such expeditions. That level of knowledge, stamina and downright determination, which might normally be admirable, yet channelled into pure destruction. A passion for ospreys, as we have seen so often with modern egg thieves, can be put wholeheartedly to bad, not good, effect.

A few of these men have found a place of their own in history. Some of the most dramatic exploits were carried out in north-west Sutherland where collectors like Charles St. John and William Dunbar made incredibly difficult journeys to collect ospreys and their eggs. They would even travel overland with a boat in order to reach the more inaccessible islands. 17 May 1848 might have been typical of their expeditions. The story goes that St. John and Dunbar went to Loch an Laig Aird just north of Scourie, looking across the loch to see the white head of the female incubating her eggs in the eyrie on a rocky islet. While his

Osprey rock on Loch an Laig Aird in north-west Sutherland as St. John would have seen it

friend went for the boat, St. John shot the female as she flew by. Soon afterwards, the male bird returned with a fish for his mate and St. John writes 'that he flew around, plainly turning his head to the right and left as if looking for her and as if in astonishment at her unwonted absence'. Soon after that, he took the two eggs from the nest and as he rowed away he recalls 'the male bird unceasingly calling and seeking for his hen. I was really sorry I had shot her'. And yet, I do not suppose that shooting was his last.

The famous nests in Strathspey were under similar siege despite, at last, the attempts of the more enlightened landowners to protect them. In 1851 Lewis Dunbar, William's brother, walked through the night to Loch an Eilein, near Aviemore, where the ospreys were nesting on their deserted castle on an island in the loch. At three o'clock on the morning of 3 May, he slipped into the icy waters and swam to the castle. It was snowing, and his climb up the castle ramparts to the nest was hampered by six inches of snow. He took the two eggs and swam to the shore, one in each hand. He blew the eggs in the boathouse, washing out the insides with whisky, no doubt celebrating his success with a dram or two himself. Like so many other eggs, these were then purchased by collectors in the south.

In those far-off days, the shooting of the adults and the robbing of the nests for the eggs and young as collectors' trophies was regarded as an acceptable activity. As the years went by, though, there was increasing concern for the protection not only of the osprey but of other species that had suffered similar levels of persecution. As the population sank to just a few pairs by the end of the century, the Grants of Rothiemurchus at Loch an Eilein and the Camerons of Lochiel at Loch Arkaig made valiant attempts to protect the birds on their land, but successful breeding was very rare. The birds reared two young at Loch an Eilein in 1896 but they failed over the next three years and the lone osprey, which returned to the eyrie in 1901 and 1902, was the last sad occupant of that famous nest. The pair at the nest on an island in Loch Arkaig, near Fort William, regularly nested until 1908 when it is last recorded that they reared young. By this time, the nest had been given barbed wire defences and placed under police protection, as it had been robbed in 1889, 1897, 1899, 1901 and 1902. Sadly, only one bird was present in 1909 and while it returned for the next four years, it was always without a mate: the end of another famous nest site. The efforts of the two landowners who had tried so hard to protect the ospreys were rewarded by the presentation of silver medals in April 1893 from the Zoological Society of London.

The history books record the last known nesting pair as being at nearby Loch Loyne in 1916. From that date on, wisdom has it that the osprey became extinct in Scotland. As William Dunbar wrote to the famous egg collector John Wooley: 'I am afraid that Mr St. John, yourself and your humble servant have finally done for the ospreys.' Were there, I wonder, any further feelings of sorrow and regret for the part he had played in their demise?

Old nest sites

Friday, 19th April 1963. To the old nest sites near Loch Morlich; redpoll singing. Male osprey soaring and displaying and at 1200 hours flew to the old eyrie in Scots pine. Walked there and found nest has been built up and a pair of ospreys were collecting sticks and re-building the nest at 12.20 a.m. Male osprey off again and displaying twice more. Two male capercaillies in pinewood and single pairs of crested tits and Scottish crossbill. Went back to the hillocks to watch the male displaying.

But were they ever extinct?

In my expert view, ospreys never did become extinct in Scotland. I remember that, in the early 1960s when I was a young warden at Loch Garten, the occasional visitor would quietly assert that a family member or friend had knowledge of ospreys nesting before the war and so the ones that we were trying so hard to protect were not, in fact, the first in 50 years. Some of these stories sounded plausible but, as one of the mistakes of the young is not to listen to older people, they were not followed up as carefully, perhaps, as they should have been. I, like everyone else at that time, was certain that the pair of ospreys at Loch Garten were the first for 40 years or more. It is important to realise that, when I first started bird watching in the early 1950s, the days of shooting a rare bird to be certain of its identity were only just on their way out. The old adage 'What's hit is history, what's missed is mystery' still, in some quarters, prevailed.

The history of ospreys in the first half of the twentieth century has always been slightly confusing. There are definite published records for breeding at Loch Arkaig in 1908 and again, successfully, at Loch Garten in 1959 and this became the period of official extinction. For a long time, too, it has been accepted that a pair nested in 1916 at Loch Loyne, not far from Loch Arkaig, although there is no published account of that pair. It is also understood that the first returning pair bred and reared two young at Loch Garten in 1954, although again, there is no formal account. Some books refer to occasional birds in the intervening period and even the possibility of breeding taking place at Loch Luichart between the wars.

About 15 years ago I decided to look carefully at records of ospreys during this period. I searched the old ornithological literature and spoke to elderly people who could remember some of the early times. When I did this I realised just how valuable it would have been to talk more carefully to local people 30 years before. By now I had the advantage over earlier colleagues, who had only been able to study records written before the 1970s, in that I knew so much more about ospreys and how they live. This was not always, after all, a shy bird of wild places and it

Sir Donald Cameron with his family's osprey medal, January 1998

was not only found nesting in remote Scots pines. We had learned, for example, that there was no regular migration route from Africa through Scotland to Scandinavia, as had earlier been believed. Many aspects of the ospreys' ecology and behaviour both at home and abroad would have come as a surprise to those early ornithologists

Looking carefully at the Loch Arkaig example, we know that a pair attempted to breed there in 1908 and that, more interestingly, a single bird, probably the male, returned each year until 1913, spending time re-building its eyrie. The loss, therefore, was not an immediate one: it was what we now expect, in that single ospreys, for as long as they survive, will continue to return to their nests in the hope that a replacement mate will arrive. On 21 September 1914 two ospreys were seen on the Beauly Firth, and this, to me, sounds suspiciously like local breeding. During the following year, between 14–20 June, one osprey was recorded as present at an unnamed loch in the Highlands, while a year later, in 1916, the pair at Loch Loyne reared young for the last time. In 1922 single birds were noted during the summer in Wester Ross, Loch Lomond and the Lake District and, in the following season, one was at a loch in the Cairngorms, while a single osprey was shot in Cumbria in the May of 1923.

Links to the past

27th January 1998. Morning went to Achnacarry to see old Donald Cameron of Locheil to ask him about the ospreys which were released on Loch Arkaig in 1929. He could remember the day – he was courting – but he could not remember much else or what happened to the birds. He had also been unable to find anything in the estate papers. Reminisced with him about the old ospreys' nest on the island and also looked at the silver medal awarded to his grandfather for protecting the ospreys in the last century. Went up to Loch Arkaig to look at the islands and talked about their history. All very interesting.

An adult male was shot on the Isle of Man on 26 March 1924, definitely suggestive of an ill-fated osprey heading for Scotland. In 1925 four ospreys, including flying young, were seen during the last week of July at Loch Riach, five miles north of New Deer in Aberdeenshire. This record was rejected as improbable by George Waterston in 1966, but from our present day knowledge of ospreys, it seems as though it was probably correct. The last week of July is, after all, just the time that a brood would be flying. The following year rumours of breeding were reported and there is also an interesting record of an osprey fishing on the River Don in Aberdeenshire on 2 June. It was reported to have caught a fish before flying off towards the north-west.

Was it flying back to its nest? We can only speculate.

In 1929 Captain Knight, a famous falconer from Kent, visited the United States to give flying displays to theatre audiences using his tame golden eagle called Mr Ramsay. Before his return on an ocean liner, he collected four young ospreys from Gardiner's Island near New York, where he had previously been photographing them. He reared them in captivity and, with the permission of the Camerons of Locheil, released them on the old nesting island at Loch Arkaig. Photographs of the young appeared in *The Scotsman* newspaper in August. I questioned Sir Donald Cameron about this in 1998 but he only remembered that Captain Knight, with others, rowed out to the small island where the nest had been and left the birds there. The nest tree had broken off in the 1920s but still had barbed wire at its base. He could not remember any birds being later found dead so we can assume that they flew off in September. Did any survive to return to the Highlands? We do not know.

Philip Brown, writing 50 years later, stated: 'Few serious ornithologists ever expected that it [the osprey] might nest again. But evidently Captain Knight did, for he released those four American birds at Loch Arkaig! The effort, however, seems to have been a complete failure and some purists might rest easily in their graves as they were, in any case, "the wrong stock".' George Waterston wrote in 1962: 'In September 1929, C.W.R. Knight released four young from Gardiner's Island at Loch Arkaig, all from different nests. The season was too far advanced and the attempt was unsuccessful; the birds disappeared and did not return.' These were birds from another sub-species of osprey (hence Philip Brown's 'wrong stock' comment) but it is just possible that one or more might have survived and returned in future years to confound the experts.

Old osprey island on Loch Arkaig, 2004

Loch Arkaig in 1929 where the ospreys from Gardener's Island were liberated

Records of sightings increased during the 1930s. An osprey was observed eating a fish by a loch in south-west Ross-shire, on 21 June 1930 and a juvenile female was killed by overhead wires at the bridge over the River Spey, near Grantown-on-Spey, on 17 September of the same year. A bird seen migrating through Cumbria on 2 May 1931 could have been heading for Scotland, while there

Captain Knight too believed that ospreys could once again nest in Scotland – did any of these 'emigrant' ospreys survive or return to Scotland?

BACK IN THE SCOTTISH HAUNTS.

In the hope of re-introducing the osprey into Scotland I brought back two pairs of the birds—two females and two males—all from different nests.

"Sea Hawks" closes with scenes of their liberation on that little island in the Scottish Loch where our story opens and where every protection will be afforded them. It is to be hoped that they will remain in that neighbourhood and that if they nest there it will be in some situation inaccessible to the collectors.

An advert from1929 for Captain Knight's presentation of his 'sea hawks'

were sightings from lowland Scotland in May and June 1932 and likewise in 1933 and 1934. A contemporary account stated that 'they might be Knight's bird or birds' so someone, at least, believed in his project. In the mid-1930s, a report by O. J. Pullen states that a few people, all pledged to secrecy, had been privileged to watch a pair of ospreys which had nested successfully for three years in succession on a certain loch in Galloway.

It was at this time that things started to hot up in Strathspey, at the very place where nesting was proved in 1954. Miss Isabella MacDonald of Inchdryne, the closest croft to Loch Garten, remembers her uncle William MacDonald remarking that the 'white birds' were back, nesting in the tree in the moss at Loch Garten, talking as though he knew the birds well. He died in 1935 so this account would relate to a period in the early 1930s. In 1997 I talked with Mrs Madge Smith, then aged 74 and living in Nethybridge. She recalled her uncle, Davie Stubbart, telling her one summer in the mid 1930s, almost certainly in 1935, when she was on holiday at the sandy beach at Loch Garten, that ospreys were nesting at the other end of the loch. Stubbart was a skilled nest finder who was paid by early bird photographers to find nests for them to photograph. He was secretive and possessive about these nests. These two contemporary accounts confirm each other and put the known occupation of the Loch Garten eyrie back by 20 years.

Desmond Nethersole-Thompson, who reported the occupation of a nest in 1954, saw an osprey over a loch in Strathspey in April 1936 and in April 1937. Nethersole-Thompson had begun studying birds in the Cairngorms but he and Stubbart did not work together, and so he knew nothing of the Loch Garten nest. In 1938 the famous naturalist, Seton Gordon, records that his wife saw a pair of ospreys during May at Loch Garten; one had a fish and landed in the tree beside its mate. They hoped that they were nesting but they regarded them as migrants on their way to their nesting grounds in Scandinavia. We now know that ospreys seen in May at Loch Garten were certainly not on migration to Norway or Sweden: they were undoubtedly living nearby.

REVIVING A SPECIES.—Two of the four ospreys brought fror America by Captain C. W. R. Knight for liberation in this country,

During World War II Loch Garten was a restricted military area for the storage of munitions and the nesting area was placed out of bounds between 1940 and 1945. In Cumbria, single birds were seen on 2 May 1940 and on 28 August 1941. In that same year, an osprey was seen near Loch Arkaig in late May and, three years later, in 1944, one lived at Loch Fad on the island of Bute between 10 June and 24 August. In 1945 a pair of ospreys was seen at Loch Ness on 24 April and when the observer returned in August, ospreys were reported once again.

In 1947 Hamish Marshall of Grantown-on-Spey saw the osprey nest at the south end of Loch Garten while he was with Davie Stubbart. While watching the birds they met Bob Grant, the Abernethy keeper, and all agreed to keep it secret. Hamish returned on fortnightly visits and remembers seeing young in the nest. He saw them nesting again at the same place during the following summer. That year, one osprey was shot in Kingussie and, in 1949, two were seen in Strathspey and a Swedish ringed bird was found dead in Moray on 22 May.

Although ospreys were being seen in the Highlands in the early 1950s, the ornithologists of the time thought that these were birds on migration through Scotland to the nesting grounds in Scandinavia and did not therefore think that they were nesting. Ospreys were seen in 1950 and 1951 in Strathspey and there is an unconfirmed report of breeding in 1951. Madge Smith, my Nethybridge contact, remembered that Davie Stubbart told her that ospreys were nesting in 1952. Hamish Gordon, an Abernethy Forest gamekeeper at that time, told me that he saw an osprey nest on top of a big Scots pine in North Abernethy Forest. In the early 1950s John Rattray of Lynamer saw ospreys at a large nest on top of a Scots pine on the banks of the River Nethy, south of Forest Lodge. Finally, in 1953, Nethersole-Thompson saw three ospreys on the wing in Strathspey on 29 August, but he failed to find the nest.

An analysis of the published records for England and Wales during the six five-year periods between 1915 and 1945 show that between one and five ospreys per year were seen on migration. Some of these, especially in the south-eastern counties, would be birds nesting on the Continent, but others clearly fit the pattern of ospreys on a regular migration to and from Scotland.

Relying on my present knowledge of ospreys, I consider that the eyrie tree at Loch Garten was in use from the early 1930s, probably with some absences, through to 1954, with young reared in some of those years. I also believe that either that pair or another built nests elsewhere in Abernethy Forest in the early 1950s, and that breeding took place in Aberdeenshire in 1925 and maybe also in 1926, and in Galloway and Loch Luichart between the wars. Ospreys nested, I believe, near Loch Ness in 1945 and possibly elsewhere. We now know that, occasionally, a Scandinavian migrant becomes disorientated in spring when it has drifted with the wind across the North Sea in bad weather. Such birds may remain here and join the Scottish breeding population. My firm belief is that occasional breeding occurred in Scotland between 1916 and 1954 and that sometimes the succession continued through a single adult returning regularly to an eyrie. Eventually, it would be joined by a Scandinavian newcomer or a Scottish-bred youngster and so breeding would continue.

Loch Garten

5
The breeding and ecology of the osprey

Spring arrival

Each spring, late in March, we start to watch for the return of the ospreys. They are one of the first heralds of the spring and summer and, for me, far more indicative of the changing season than any cuckoo or swallow. Most ospreys return to Scotland in the first ten days of April but, in the last decade, a few very early birds have come back in mid March. At the other end of the scale, some late adults and young first-time breeders may not arrive until late April. Ospreys do not arrive as pairs because they do not winter together in Africa. If both partners have survived, they usually appear within a week or so of each other and pair up again. The male is usually the first to be seen at the old nest but, sometimes, the female beats him to it.

Between 1991 and 2005, the mean dates of return at the well-watched Loch Garten nest have been: 2 April for males (with a range of 27 March–9 April) and 11 April for females (with a range of 27 March–27 April). In previous decades, the equivalent dates for males were 9 April (with a range of 27 March–22 April) and for females, 10 April (with a range of 28 March–26 April). In 22 of those years, the male arrived before the female, with the situation reversed in 12 seasons during that period. Both partners have arrived on the same day in four breeding years.

Ospreys clearly set off from Africa with the intention of nesting at the same eyrie that they used the previous year and the chances are high that both will return to breed again with the same mate. Our studies show that the annual mortality of mature ospreys in Scotland is about ten per cent, meaning that nine out of every ten adult ospreys leaving Scotland in the autumn will return to breed in the following spring. Changes of mate are unusual unless one of the pair has died but if a bird is delayed its mate may pair up with a new bird. Fights break out when the late bird eventually arrives and tries, not always successfully, to oust the usurper.

When migration has been very strenuous, some ospreys arrive exhausted. It is a hard time for the birds but if the weather is kind, they soon regain their strength. If it is a cold late spring, though, with snow and ice or heavy rain and winds, fishing is difficult and they find it a struggle to re-build their strength. Some may even die in the last days of March and early April. In April 2007 the weather was sunny and warm in Scotland, but in Spain and the Bay of Biscay it was very unsettled with spells of high winds, overcast and rain. Nearly twice as many adults as usual failed to get back to Scotland and breed, and others were very delayed.

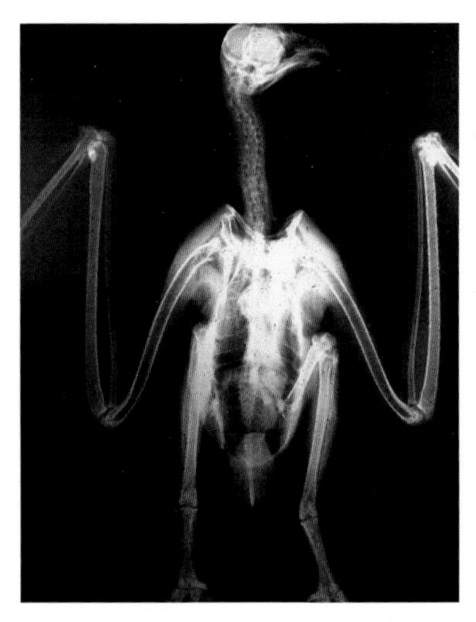

X-ray of an osprey which had died of starvation having been too weak to catch fish after a difficult migration

I remember, in 1975, receiving a call from a gamekeeper in Nairnshire. He had been walking along the River Findhorn and had seen an osprey perched in a tree but, while he expected the bird to fly off when he approached, it had instead glided down and fallen into the river. He waded out to rescue the exhausted creature, still alive but very weak. He took it home and put it in a box by the cooker to warm it up, before going out to catch a couple of small trout to feed it. In the morning, though, he called me back to say that the bird had died. When I went to collect it, I found it was a male, in a terribly thin and emaciated state, with a leg ring identifying it as a chick I had ringed six years before at Loch Garten. I remember that I called in to Raigmore Hospital in Inverness on the way home to ask if they would x-ray the body to check for shotgun pellets and, instead of sending me off to the vet, they hurried me to the front of the queue. The resulting x-rays showed that the poor bird had clearly endured a very difficult migration, perhaps becoming lost at sea, and, once back in Scotland, had simply been too weak to catch fish. It had died from malnutrition and our recent satellite studies show that this is an ongoing hazard for returning ospreys.

Courtship

In the early days, during the 1960s and 1970s, when the species was rare, we were treated to the enthralling sight of male ospreys performing the most beautiful and dramatic aerial displays over their nesting grounds. In full display the male would return to the area near his nest carrying a fish, flying very high up, possibly at an altitude of one thousand feet. He would fly in great sweeping circles with the fish perfectly aligned in his talons, clearly displaying his prowess for the benefit of all other ospreys but for the females in particular. In fast flapping flight he would call with a distinctive and very high pitched 'pee-pee-pee-pee', a sound many people found hard to believe could emanate from this large bird of prey. Suddenly, he would fall steeply for a hundred feet or more and then pull out of the dive before rising, with rapidly beating wings, to regain

Osprey's display

14th April 1960. Loch Garten. 06.55 a.m. Male arrived with fish, started to display, continuous calling 'cree-cree-cree'. Wings flexed, beating faster than in usual flight, tail fanned, and the whole time floating up and down like a puppet on a string. Occasionally circling at 300–600 feet above ground. 07.05 a.m. drifted away to the east. 11.15 a.m. short display and sailed in from the east with a half eaten fish, soared over the nest.

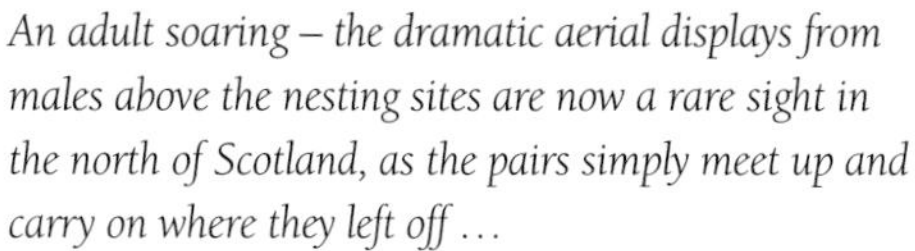

An adult soaring – the dramatic aerial displays from males above the nesting sites are now a rare sight in the north of Scotland, as the pairs simply meet up and carry on where they left off …

Diving osprey displaying his prowess

his previous level. There he would hover for a few minutes, wings working at full speed and tail fully spread, before diving once again. He would carry on this roller coaster display all around his nesting area, and the up-and-down flight, with the high pitched calling, would continue for anything from ten minutes to half an hour. At the end, he would close his wings and dive towards the nest in a series of spectacular swoops and checks. Should his mate be in residence on the nest, he would fly straight down, land beside her and present her with the fish. If he were still awaiting her return or trying to attract a mate, he would stand expectantly on the nest with his fish, hopefully scanning the skies.

Now that the osprey population in the north of Scotland has grown so much, we rarely witness these breath-taking displays there, although they do continue in the new nesting areas further south. At the well-established nests, where we expect to find the same birds breeding year after year, the pair merely arrives at their nest, meet up and carry on where they left off, as though last August were only yesterday. The birds start to repair their nest and make good any damage from the previous winter, although there is often very little to do at the large eyries. The male then starts going off to hunt and bring back fish for the female who, now that she has returned and settled in, will not leave the nesting site until the end of summer. If she is first home, then, of course, she has to go and fish for herself until her old mate returns or she attracts a new partner.

If the male is waiting for his mate or is trying to attract a new one, or is, perhaps, a young bird trying to attract his first partner, he will spend a lot of time near the eyrie. I have often watched a male perched like a sentinel on a favourite branch in a prominent tree. His white head gleams in the sun and he holds his wings slightly open to expose the white feathers between the wings and his body. These big white flashes are very obvious to any flying osprey, pulling in females and warning off males, advertising that this is his territory. And then, suddenly, the stillness and grandeur will disappear in a flurry of nest-building, as he frantically prepares for that still-elusive female.

During the days before egg-laying, those birds which are in pairs will often perch together in a favourite tree. They can look very lovely and even romantic together, their outlines against a dark sky, the black of distant snow clouds bringing out the white of their feathers. During this period, mating is frequent and it nearly always takes place on the eyrie, often interspersed with nest building. The male lands carefully on the female's back and during successful copulation his tail is bent down and wrapped around that of the female. His talons are closed as he balances on her back. Copulation is often unsuccessful and the birds distract themselves by flying off to find more nest material. This behaviour continues for about two or three weeks until egg-laying takes place and there are in the order of 160–190 copulations during that time, of which about 40% are successful. The number of copulations reaches a peak about four to eight days before the first egg is laid. Sometimes, an extra male may attempt to copulate with the female, especially when the male is delayed, so not every young in a nest has the same father.

A male osprey with his large catch

The male must, at this time, supply plenty of fresh fish to bring the female into good condition after her migration. If he is a good hunter he will bring two or three fish each day, often heading straight to the nest with no delay. When a male osprey heads off on a fishing trip, he does so in a most distinctive way, with a determined flight that I have learned to recognise. It is so impressive in a humble way, as though the bird knows that catching fish for his mate or growing family is the most important duty in the world for him. It is indeed 'second nature'.

Nests and trees

The female may be the one to stay around the nest but both birds are active at nest-building. Most eyries, especially larger well-established ones, survive the winter, but they are sometimes destroyed by very severe weather. The birds search their local surroundings for dead branches and other suitable material and can go to considerable trouble to bring back a piece that they consider worthwhile, diving down in flight to break off the end of a dead branch, maybe several feet long, before lumbering back up to the nest. Males tend to bring in the largest sticks, anything in length between 15 centimetres and one metre, while the females usually fetch more in the way of grass, moss and bark. Males generally build more often early in the season, while females are more active later, when the male concentrates on hunting. It can take a long time for the birds to push and pull the branches securely into position although they can be surprisingly industrious, once they get started. In 1961, for example, the male at Loch Garten brought in 337 items and the female 77, all in the period 10–28 April.

Active nest building

8th April 1960. Osprey back for first time, only base of nest left from last year when we visited the tree, plenty of sticks under the tree. 9th April. 10.32–11.30 a.m. Fourteen sticks taken to the nest by the male; he spent a lot of time pushing sticks into place and arranging them with frequent breaks to look hopefully into the sky. Every stick collected in flight, on one sortie it took him three swoops to get the dead stick to break from the tree. Once collected a stick too heavy to carry which fell but only one stick fell from the nest after he took it there.

Some pairs, even though they already have a large nest, will keep on building and making it bigger when, to human eyes, there seems to be little need. Most nests are constructed on tree tops and are about 30 to 40 centimetres tall and a metre or more in diameter but after years of building, some may be a metre high and are then vulnerable to strong winds. Nests on cliffs, on man-made structures or on the ground can grow even larger and can reach 1.5 to 2 metres in width and more than a metre in height. Their weight is immense, by any standards, anything up to 200 or even 300 kilograms. Eyries, even in trees, can last for many decades while those on rock pinnacles and small islands may be used for hundreds, or even thousands, of years.

But not all ospreys are hard workers. Some, when they return in spring, just scrape away the middle of the nest to make a hollow, lining it with dry grass and mosses and poking a few new sticks in around the edge. I have watched many male ospreys scratch out the bottom of the nest, scraping away just like a lapwing in a field, body well down, tail sticking up, earth flying out, making a scrape for the new eggs. It might not look quite worthy of such a noble bird, but it is a job worth doing as once the nest is lined they are ready to breed. By the end of the season, the nests will appear flat and are often quite earthy on the top. Sometimes they are even bright green with new grass sprouting, because a female adds a lot of soft material while the young are growing, in order to create a flat platform for when her chicks are older.

During the 1950s and 1960s ospreys were only associated with Scots pines in remote parts of the Scottish Highlands. Nowadays, though, they live in many types of habitat, from those traditional Scots pine forests to lowland farms, in open woodlands or plantations, near rivers, lochs and estuaries, and even on open coasts. In fact, as long as there are fish to be caught and suitable nest sites, ospreys can live almost anywhere, even in close proximity to people. Nest trees may be overhanging the water but in most cases they are some kilometres away from the fishing grounds. Nests are usually spaced at a couple of kilometres apart in Scotland but, as numbers

Being shown a new nest

18th April 1971. Nest BO3. Harvey and I left 6.15 a.m. and drove to meet Donald Frazer the woodcutter who took us to see a new nest where he is felling trees. We walked from the road. Female osprey flew off the eyrie and circled. About 30 ft tree – live Scots pine in clearing about hundred yards from edge of wood. Some dead trees. Tree easy to climb, nest right in the crown – well built – several years old. No eggs – nest nearly ready. Saw a few capercaillies and siskins. Walked back to the car and drove home.
Later report: local youths up the tree 25th April and they say one egg. Went to nest on 27th, female sitting and did not move, in fact very quiet until we walked halfway to the eyrie. Male on a dead tree also quiet. Then both up and circled us closely – two eggs in the nest. We sawed off all the lower branches and left in 15 minutes. Female back onto the eggs when we only halfway back to the car. All satisfactory.

have increased, so has density and the distance between nests may now be only a few hundred metres in some areas. Exceptionally, in a few places like Florida, ospreys may nest even closer together: ten nests within an area of one hundred square metres is not unheard of.

Scottish ospreys continue to prefer conifers for their nests because they provide strong broad tops and a secure base for building. However, even in the Highlands of Scotland, the types of trees used in recent years have varied quite considerably: live Scots pine (53), dead Scots pine (12), Douglas fir (21), silver birch (3), alder (2) and single nests in redwood, oak and wild cherry. Ospreys in Scotland, as in other countries, have also built nests on tall electricity pylons. To date, though, there is little sign in Britain of their nesting on deserted buildings, rocks and cliff faces, as they were known to do long ago.

Large bulky osprey eyries also provide homes for other creatures during the summer. Several nests I know have had jackdaws nesting in them, much to the annoyance of the ospreys, while another nest in Moray was home to tree sparrows. Pied wagtails have also used them, while in North America, house sparrows, purple grackles, house wrens and even night herons have been seen in residence. Pine martens also burrow out holes in the bottom of nests as safe dens. At Rutland Water, Canada and Egyptian geese have laid eggs in osprey nests and, elsewhere in the world, old nests have been used for nesting by bald eagle, great horned owl and peregrine.

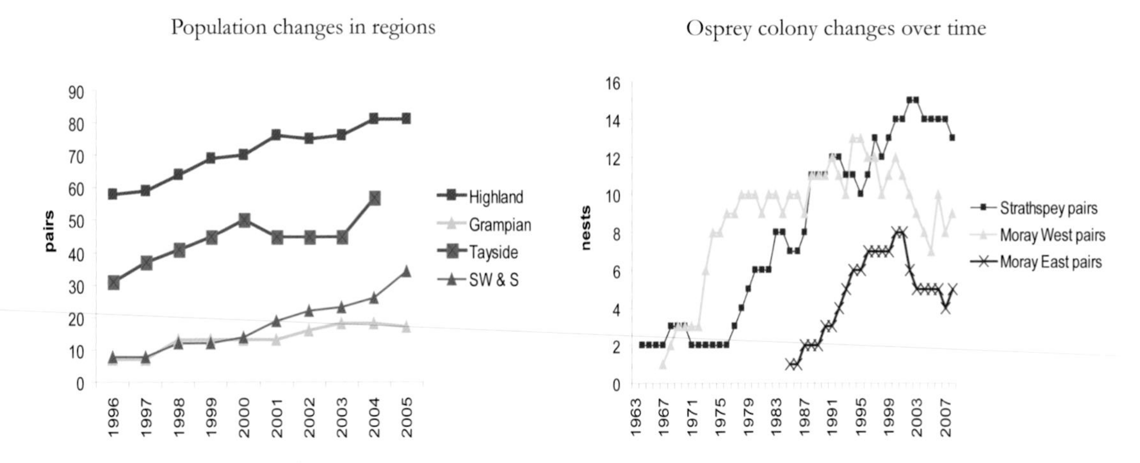

Eggs

A mature female osprey usually lays three eggs at two-day intervals so that it takes five days to complete the full clutch. Very rarely in Scotland we find four eggs laid, but this only happens when the female is in exceptionally good condition. Young females breeding for the first time at three years of age are more likely to produce two eggs in their first clutch, only laying three in later years.

The female starts to incubate after laying her first egg. For the human watcher, the best proof that she has laid her first egg is when she is observed to be incubating through the night. The earliest date that we have recorded for the first egg is 10 April but eggs are mostly laid from the middle of April through to the first days of May. Late pairs with young females may lay their eggs later in May, the latest recorded being 23 May. Elsewhere, eggs are laid as early as mid March in the Mediterranean and as late as mid-June in Lapland.

Osprey eggs are beautifully coloured and patterned. They are the size of chicken eggs, averaging about 62 millimetres in length and 46 millimetres in breadth. The background colour varies from buff to stone to nearly off-white and the egg is patterned with chocolate, mahogany or ochre, the colour most commonly seen being a dark, rich brown. The eggs are patterned with blotches, spots and streaks, with most of the markings being on the large end of the egg. The patterns and colours of the eggs can vary within the same clutch or be remarkably similar. Once I have seen a pure white egg in a clutch of three. We know that the eggs laid by any particular female are similar from year to year, but they differ from those of other females. In fact, in situations where we do not know the actual identity of the laying bird, the coloration and patterning of the eggs gives us a good idea as to whether this is the same female as in the previous year or whether there has been a change.

As part of our studies, we are keen to record the number of eggs in each nest in order to keep a check on the breeding success and productivity of each breeding pair. We use a very long aluminium pole with twelve sections, each five feet in length, with a small mirror fitted on the end. By using binoculars, we can look in the mirror to check the contents of most of the nests without unduly disturbing the birds. We borrowed the idea from the then Ministry of Agriculture pest control officers, who used similar poles to poke grey squirrels from their dreys, and while it might look primitive, it works. The checks are carried out under licence from the government and take great care to make sure that the birds are not unnecessarily disturbed or treated

About the size of chicken eggs, osprey eggs are beautifully patterned, and though they differ from other female ospreys' eggs, it is possible to identify a particular female's clutch by observing the coloration and pattern of the eggs

incorrectly. We do not, for example, visit the nests when it is raining or very windy or while a female is actually laying.

In the early days we were too worried about causing disturbance to check the Loch Garten nest, but we later learned that ospreys, if treated carefully, are very good parents and will never desert their eggs. We always walk slowly towards the nest, in full view of the birds, so that they know we are coming and have time to leave the nest carefully without damaging the eggs. They will start giving out their alarm calls 'kew-kew-kew' and will circle overhead, still calling, while the nest is being checked. As soon as we have counted the eggs and begun to walk away, the birds 'escort' us off their territory. That said, ospreys do have different, highly individual temperaments, with some becoming very annoyed and circling noisily, while others keep well away and make very few cries.

The birds are often back on the nest before we have reached our vehicle and in fact, over the years, some have become very accustomed to us and tolerant of our presence. They may only fly off the nest at the last possible moment and return to incubate when we are less than a hundred yards away.

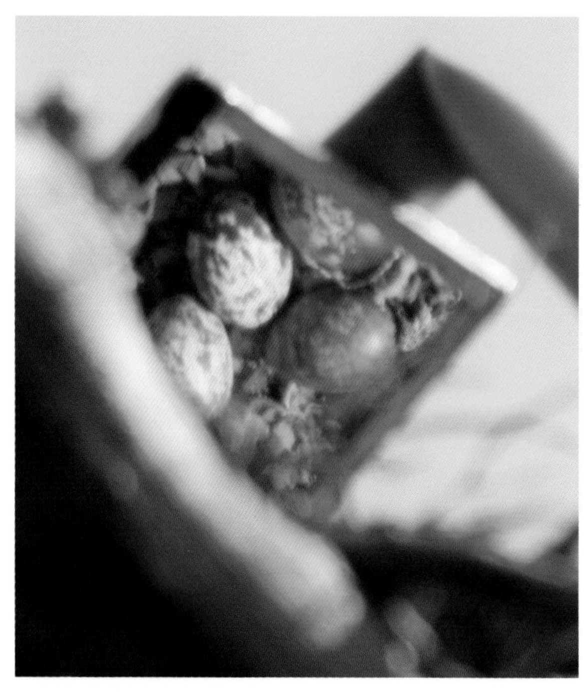

My first four egg clutch

5th May 1977. Nest A02. Went to the nest at 6 a.m. female on the eggs and male in next tree. Climbed up to the nest and amazed to find 4 eggs. Marked them '77 – A2' with invisible ink pen. Measured them 66 x 46.4 mm, 67.5 x 44.8 mm, 66.6 x 47.2 mm and 66.8 x 46.6 mm. Left and female returned to incubate. Later date, 1630 to nest; female on edge of nest; male turned up with fish. Looked into the nest with a mirror pole – no eggs – had been robbed. Obvious signs of climbing up side of tree. Marks on barbed wire etc. and broken ends to branches etc. Very disappointed.
The female called three times and then perched on tree K.

Incubation

During the period of incubation, life at the eyrie is usually quiet. The incubating female may be panting in the spring sun or huddled down, trying to keep her eggs warm while being lashed by rain or cold snow showers. I often think of these poor birds when I turn in for the night, hearing a howling gale or rain lashing the nest trees. The female sits throughout the night and carries out the majority of incubation duty. Her mate takes his turn for anything from 20 to 35% of the daylight hours, with some males keen to incubate and others slow to take over this particular parental duty. A male may sit for anything from thirty minutes to a couple of hours, with the longest periods usually occurring after his morning fishing expedition.

The main task for the male osprey is to catch fish for the female and for himself and so he is often away fishing very early in the morning, usually by seven in Scotland. More often than not, he will fish again in the evening. Often the impetus to go and fish stems from his mate's peevish nagging calls to him, sometimes lasting for half an hour or more. This querulous call 'quee-quee-quee' means that she is hungry after a long session on the nest.

During the incubation period, the birds exhibit a typical daily pattern of behaviour. The male returns from fishing and flies to a favourite perch to eat the front half of his catch. Ospreys always start with the head, leaving indigestible parts like large jawbones, stringy guts and sometimes even tails to be dropped to the ground. After feeding for about half an hour, he will fly to the nest with the remainder of the fish. He lands gently on the edge of the nest and by this time, the female has carefully stood up from the eggs and gingerly walked towards him, often with her claws withdrawn. Without further ceremony, she drags the fish from under his feet, although sometimes he can be rather reluctant to hand it over! She then flies with it to one of her favourite perches while her mate moves carefully forwards to settle on the eggs and begin his stint of incubation.

The ravenous female often starts, quite ferociously, pulling the fish to pieces and only slowing down when her first hunger is satisfied. Often she does not stop until the entire fish is consumed, ending up by swallowing the tail. Next she will thoroughly clean her feet and bill and often start preening her plumage. This may take more than half an hour, and it is not hard to imagine how pleasurable this break from the nest must be. She may fly around to get some exercise, perhaps chasing away the local crows if they have strayed too close. If the nest is near a loch she may trail her feet in the water to clean them of fishy debris. After perhaps an hour has passed, she will return to the nest to relieve the male of his duties. Some males leave immediately as soon as their mate returns but others continue to sit. The lucky female may then go off again for a short period but after ten minutes or so she is usually back, quite often returning with fresh nesting material for the eyrie. At times, it even looks as though she pushes her mate out of the nest in order to continue her incubation.

Female feeding

When the male leaves the nest, he flies to a favourite perch that, quite often, is hidden out of sight. Ospreys nearly always perch on just one leg with the other leg withdrawn up into the feathers; an unhelpful habit for anyone trying to read ring numbers through a telescope. In perching, three toes point forwards and one back and the bird looks very comfortable. Occasionally, he may stretch back and look around but sometimes he actually closes his eyes for a brief moment and can nearly fall from the perch. Just as you are trying not to laugh, though, he will spring immediately to attention, before a potential threat appears. The male will sometimes bother to chase crows but will certainly and determinedly drive off any more threatening predator, such as a passing peregrine or golden eagle. He pursues them with great gusto, not giving up until he has chased the intruder out of the nesting area. He will likewise attack any intruding osprey of either sex, although the truly aggressive attacks are reserved for males. These episodes can be extremely stressful for the incubating mate. Finally, at the close of the day, the male flies off again to catch another fish, with just over two a day being the average number brought back to the nest during the five-week incubation period.

Just hatched

Other troubles may be in store during incubation and the most heartbreaking is a raid by human egg thieves during the hours of darkness or early morning. Natural hazards include the breaking of eggs when two females fight over a nest and occasional raids by pine martens, which have become more frequent with the recovery of this species. Strong winds and rain pose a particular threat, especially for young osprey parents with new nests, and many an eyrie has been blown out and destroyed during very high winds. After any dramatic egg loss, most pairs lose interest in their old nest and may start building a 'frustration' nest in the vicinity. Only exceptionally will they replace their clutch but they will often use the new nest during the following season. At other times, eggs may fail to hatch due to infertility or because the birds have suffered undue disturbance and in these circumstances, the pair tend to stay around that nest until the time comes to make an early departure for Africa.

Moulting

When you look carefully at adult ospreys through a telescope, you can see that the mantle (the back) and wings are not a uniform dark brown, but are made up of fresh, dark brown feathers among pale washed-out ones. This mixture is due to the moult, a process that takes two years to complete. Incubation is a good time for female ospreys to renew their worn feathers but the males, of course, have to be in peak flying condition during this time and so their moult occurs in winter. The large, strong feathers of the wing and tail last for two years before being replaced, unlike in most smaller birds which moult once or even twice a year. Some of the feathers are moulted in Africa and their replacements look fresh and new. Others, which have suffered that extra year's wear and tear in the searing sun of Africa, appear ragged and very worn. The moult of the big flight feathers of the wings is quite unusual in ospreys, in that it starts with the innermost feathers and occurs in successive waves across the wing. This variation in moulting times can be seen very clearly when you examine the range of light and dark feathers in an osprey's wing or tail.

Young in nest alongside dud egg

Ospreys spend much time preening and oiling their feathers from the oil gland above the base of the tail. It is essential to keep their plumage in good order. They also bathe by standing in shallow water and ducking their head and breast in the water, although in Scotland they usually wash their feathers in the rain. They clean their bills carefully after eating, and sometimes wash their feet by dragging them in the water as they fly.

Fishing

Ospreys are specialised fish feeders and over all my years of personal observation, I have never seen them eat anything else. There are, though, very occasional records of their taking water birds or small mammals as prey. The feet are specially adapted, having very long, sharp, curved black talons and

Wing moult

toes that are covered in spiky scales (spicules) for grasping a wriggling fish. They also possess a special ability to reverse the outer toe so that instead of three toes pointing forward, as in a normal bird, the osprey is able to change its grip to two forward and two back. This helps greatly with catching and holding the prey. The feet have a very powerful explosive grip when catching a fish, and it is alleged that it takes only one fiftieth of a second for the feet to close around the catch.

Ospreys usually visit fishing grounds within 15 kilometres of their nests but, occasionally, the birds may range as far as 28 kilometres (17 miles) in search of their prey. In the Scottish Highlands, ospreys fish on all sizes of river, lochs, lochans and even, very rarely, on small ponds,

Preening feathers on a favourite roost

Defecating after preening

including those in gardens. Fish farms and fisheries are also favoured hunting grounds. The birds sometimes hunt from a prominent perch above the water, launching themselves in free fall at the fish. Mostly, they scan the surface while in flight. On arrival at the fishing site, the osprey circles the water, usually at a height of one hundred feet or more. It flies steadily above the surface, with the head swivelling and looking intently at the water. When a fish is spotted, it stops and hovers, an expert manoeuvre performed with the head pointing downwards. The bird might drop halfway to the water and continue to look at the fish. If satisfied, it closes its wings and carries out a most spectacular plunge into the water, at the last moment throwing its outstretched talons directly in front of its head to grab the fish. Very strong winds and heavy rain, as well as flooded turbid rivers, make fishing much more difficult. When hunting, they seem impervious to mobbing by other species, such as black-headed gulls and jackdaws.

Foot and talons – a special trait of the osprey is its reversible outer toe

Occasionally, the bird is completely submerged and has to haul itself out of the water, its long wings strongly beating. If the fish has been caught it is grasped tightly, often in only one of the osprey's feet. As the bird leaves the water, the other foot is brought into play and locked onto the fish to hold it, the head facing forwards. At a height of about ten feet or so, the osprey gives itself a highly distinctive shake, in order to shed as much water as possible from its plumage. Experienced ospreys are successful in one out of every four dives but others may make many attempts before landing a fish. Our observations have shown that at the best fishing sites there is a hierarchy among the local males, with the dominant ones taking precedence.

Brown trout – traditional fare for Scottish ospreys

Ospreys eat any available species of fish but prefer those in the 150–300 grams (½–¾ lb) weight range. Prey, though, can weigh anything from 50 to 500 grams and sometimes even up to 1200 grams (2½ lbs). Each osprey needs about 300–400 grams of fish per day, but some prey species have greater calorific value and are therefore more valuable and preferred. The season and weather conditions will also influence which species are caught. On warm spring days, pike come into the shallows to spawn and feed and this is the best time for ospreys to catch them. On still days, brown trout rise to the surface to catch flies and are easier to take. Once ospreys started to feed in coastal waters in the Highlands (and now elsewhere in the United Kingdom), they were able to exploit different prey species, such as flounders, which are most easily caught as the tide ebbs and flows twice a day. Ospreys learn to follow the tide, watching for the distinctive movement of flounders, and, amazingly, can take them from just a few inches of water without hurting themselves. In these coastal environments, the birds also catch sea trout as they enter the rivers while in warmer waters the favourite prey is grey mullet. Mullet is one of the most important fish for ospreys, both in coastal Europe and in Africa – after all, the osprey was once called 'the mullet hawk' in southern England, where there is a much greater choice of prey species. In such rich feeding areas, they have been recorded eating perch, roach, bream, grayling, tench and even goldfish from garden pools. In recent years, nest cameras have improved the identification of fish and researchers have detailed the fish species for several parts of Scotland, the Lake District and Rutland Water (see diagrams).

In Strathspey the main prey species was formerly brown trout or pike, but this is now supplemented with rainbow trout, mostly provided as easy targets by fish farms. During the 1970s and 1980s, I remember the early fish farmers in Scotland calling to tell me when they had seen an osprey, and, occasionally, how wonderful it was to see it soaring and then plunging into the water to catch a fish. After a few days or weeks of regular visits, though, their enthusiasm for

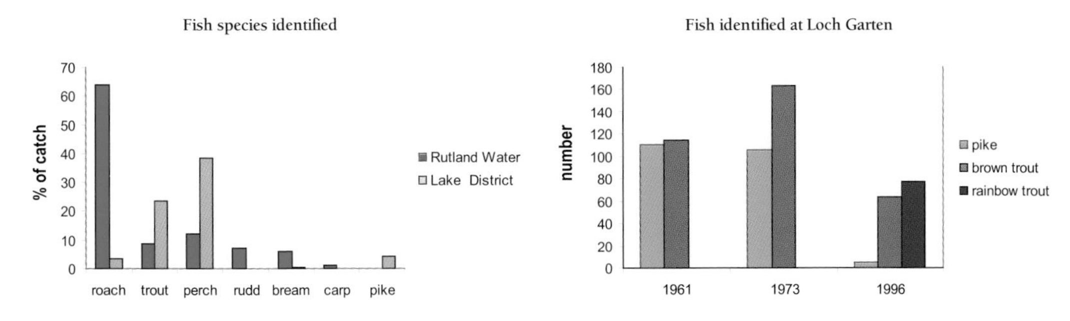

ospreys would start to wear thin, and they would then be asking what they could do to keep the birds away, or would the RSPB pay for the fish? My advice was always that ponds should be covered with netting, particularly before they were stocked with fish. Ospreys catching rainbow trout from waters stocked for angling is a much more difficult problem to solve. For those who do not share my view that ospreys are just a part of nature, losses can be minimised by stocking well before the ospreys return for the breeding season, allowing the fish time to learn their local environment and to avoid predators. Stocking with heavier fish, too big for ospreys to catch, will also help.

Once they have caught their fish, the birds are efficient feeders and discard very little as waste. On average, a pair of nesting ospreys will eat two to two-and-a-half fish each day during incubation and, as their chicks grow, so do their requirements. Broods of three chicks require seven or eight fish a day while smaller ones need fewer than six. Fish deliveries reach a peak when the chicks are about four to five weeks old, their period of maximum growth. There is then a decrease in the number of fish brought in by the male and some experts suggest that this acts as an incentive for the young to leave the nest. I think it occurs because the young have now completed their main phase of growth and it is helpful for them to be a little leaner when they take their first flight. Later, in the post-fledging period, the young put on more weight as fat for tiding them over during their long and energy-sapping migration.

Fortunately, for the success of osprey conservation, the birds do not prey upon salmon. It is not a matter of taste, but of size. When young salmon go to the sea as smolts, they are very small and not worth catching, then, when they eventually return to the rivers to breed, they are far too big and heavy to lift out of the water. Occasionally, when an osprey catches a very large fish, it may drag it ashore through the water and eat some of it on the ground before carrying it away. More usually, though, ospreys will avoid large prey or release such a catch after the first struggles of the fish show them that they have taken on more than they can handle with ease. Reports that ospreys may drown by being dragged down by a large fish, their talons embedded in its flesh, are simply not true.

I do remember one occasion, though, when an osprey ate a salmon. Sir William Gordon Cumming, one day in early spring, was casting his line on a beat on the lower Findhorn River. It was during the first days of April, and the weather was very cold, but he landed a salmon and laid it on the shingle bank. He then walked up river,

Fishing in estuaries

20th April 1977. Loch Fleet Sutherland. 08.04 a.m. Male osprey hovering at 120 feet then very fast dive into one of the channels at low tide and caught a flounder about 9 inches long. It turned and flew inland pursued by two shelduck and up into the Mound National Nature Reserve and landed in a dead tree. Left it eating fish.

casting his fly over a long pool. He walked a hundred metres or so upriver, turned, and saw an osprey perched on the salmon, tearing away at its flesh. Fortunately, since ospreys always feed from the head first, the fisherman reported relatively little damage to his catch. This osprey was no doubt very hungry and had failed to find fish of its own. We have recorded ospreys eating dead fish from stocked ponds, but this is rare and the birds will always prefer to take live prey. One incredible bird would even drop into plastic bins holding discarded fish at a fish farm and grab dead and even rotten fish.

Ospreys at Rothiemurchus fish farm

22nd August 1995. Got to fish farm at 5.15 a.m. and one osprey already there in the dusk. Set out the fish and several ospreys arrived and very busy from 6 to 7 a.m. with up to six ospreys most of the time and a maximum of nine in the area at once. Loch Garten male with the radio came in at 5.50 a.m. and caught fish and away very quickly. It returned at 6.52 a.m. and caught fish quickly and away again. Osprey '394' caught nothing. Saw one osprey chasing another with a fish swinging down – after it like a kite but gave up. Caught Norwegian male at 6.15 a.m., another female at 6.40 a.m. and the young male 'orange SB' at 7.10 a.m. Successful morning, left at 7.30 a.m.

Hatching

After 35–37 days of incubation, an air of expectancy hangs over the osprey nest. The adults will have turned the eggs regularly throughout incubation, so they will recognise any subtle changes, and already will have heard their chicks calling and turning inside the shells. When the time comes to emerge, the young osprey, like any other chick, must chip its own way out. It has a small, sharp, white temporary tooth on top of the bill which it uses to crack the shell, rotating inside the egg to allow itself to struggle free. The mother osprey is very attentive at this time, caring for the first chick while keeping the other two eggs warm, so that they too may hatch in the following days. Each chick is covered with pale brown and buff down, with a small black mask around its eyes and a pink mouth.

The behaviour at the nest changes dramatically as the chicks hatch. The female is loath to leave the nest and the male bird brings the whole of his catch directly to the nest without delay, instead of first stopping to eat some of it himself. The female tears off very small pieces of fish and feeds them to her tiny chicks. During the first week or two, the distant osprey watcher with his telescope cannot tell how many chicks are present but, frustratingly, can only see the mother bird reaching low down into the depths of the nest with tiny pieces of food. She eats all the larger and difficult pieces herself but is assiduous in her attention to her offspring.

The growing family

It is quite amazing how quickly the chicks grow, soon undergoing their first moult to don their second, grey-coloured down. The first of the true feathers start to grow when they are about three weeks of age and by then they are already starting to look like their parents, although the youngsters have orange eyes, much darker at this stage than the lemon-yellow eyes of the adults. There is quite often a difference in size between the first- and last-hatched chicks.

The male increases the frequency of his fishing trips now that the chicks are hatched and has to work hard as, by the time they are a month old, his offspring are consuming about six fish a day between them. It is rare to see any sibling aggression and, nearly always, the whole brood thrives. Aggression and starvation can occur during extended periods of adverse weather or when fishing is poor and then the youngest chick will be the first to die. In the worst cases, the middle chick as well or even the whole brood may die from starvation and cold.

When not being fed, the young lie down quietly. They might watch their parents or other birds flying over, or perhaps contemplate the bluebottle flies attracted to the smelly, fishy odour that pervades the nest. The parents, in fact, keep the nest very clean, either eating up the remains of fish or removing them to a perch from where the debris can be dropped to the ground. The female also continues to bring in fresh bedding material, especially grass and moss, to fill up the nest and to create a flat platform for the young.

There are usually scavengers around osprey nests and a pair of crows will nearly always be nesting nearby, the first to come and search for scraps. During the night, a local fox will often explore the ground under the eyrie looking for leftovers, however small.

The young birds spend a lot of time preening. It is important that they tease out the vanes of their feathers to clean off the down and waxy feather casings. The feathers, which soon will be crucial to the birds' survival, must be scrupulously cared for.

The chicks are ravenous and the hard-working father may now need to catch an extra fish to eat himself at the fishing grounds, before resuming the seemingly never-ending task of bringing food to his rapidly growing family. Maternal ospreys are probably one of nature's most conscientious parents. They continue to feed their young even when they are fully grown. The female osprey is always pushing the chicks to eat more fish, super-mothers giving constant encouragement to the chicks to eat. Now that we understand more about the migration of the young birds and that first, difficult, long distance flight over the ocean and the desert that faces them, we can understand why the mother birds are so keen for their young to be as fat as possible before those marathon journeys begin.

As the chicks become bigger, they naturally take up more room in the nest. There is more activity now and much more flapping of young wings. One chick will walk across the nest, raise itself up and start stretching, another will follow suit and soon they are beating each other round the head for lack of space. Next, a kind of truce may be called before the youngest chick will get up and start flapping too. No wonder that at this stage the female is more likely to leave the nest and perch in a tree nearby to get some peace.

She is, though, still on the watch for predators, ever ready to chase off buzzards and crows or, in rare cases, white-tailed sea eagles or goshawks which could pose a threat to her young. Their

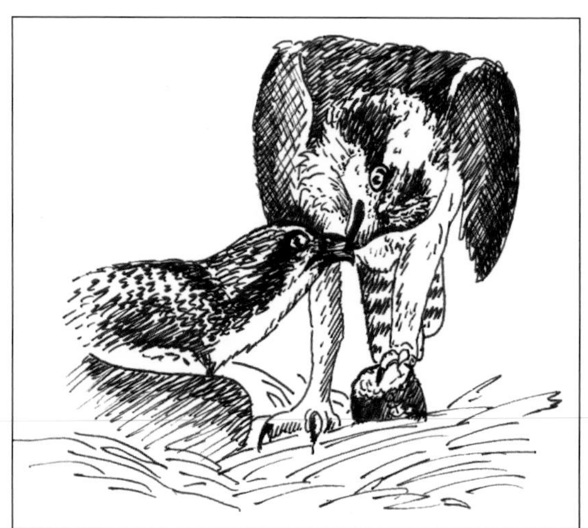

Female feeding young

Good brood of three young

mother's first alarm cries make the young lie flat, quiet and unmoving in the bottom of the nest, camouflaged by their cryptic colour patterns of brown and buff. When the danger has passed, the female flies back to the nest and the reassured youngsters start to move around again. Life returns to normal until the male brings in a fish and the feeding performance resumes.

At about seven weeks of age, young ospreys start to lift off the nest while flapping, and it is the lighter male chicks that usually manage this first, often a few days ahead of their sisters. At first they rise on beating wings for just a few inches but this soon becomes a few feet above the nest. They may rise even higher but every time the young birds collapse back into the nest, holding on grimly in panic. Just a few days later, though, they become truly free, making their maiden flight when they are about 50–56 days old. The average age for this is about 53 days and, once again, the lighter males tend to fly before their sisters. (Young ospreys in southern latitudes take longer to fledge and so the timescales here are extended). The first flight may take only a few minutes, with the bird landing back on the nest or in a nearby tree, but other flights are more adventurous and can last for ten minutes or more. In no time at all, it seems, the young have joined their parents in the skies and are actively taking flight

Juvenile feathers

Female guarding nest

around the local area, returning nonetheless to the nest as soon as there is any indication that their father is arriving with a fish. Even at this stage, the female still takes charge of the fish and often feeds her young, even though they are now perfectly capable of helping themselves.

Behaviour after fledging

As their feet and talons grow stronger, the young ospreys start to tear up fish for themselves at the nest or take pieces of fish away to eat at a perch in a nearby tree. The young find it difficult to hold on to a branch and a fish at the same time, let alone begin to tear it up and feed themselves, but, just like a child learning to ride a bicycle, the tumbles and mishaps are soon a thing of the past and it has all become second nature. By now the youngsters are as big as their parents, in fact they are even slightly larger as young birds of prey often have longer wing and tail feathers, giving them more lift in flight to counteract their inexperience. Distinctively, all the feathers in the upper plumage of the young osprey are fringed with creamy white so that they appear mottled

Sketch of female protecting nest

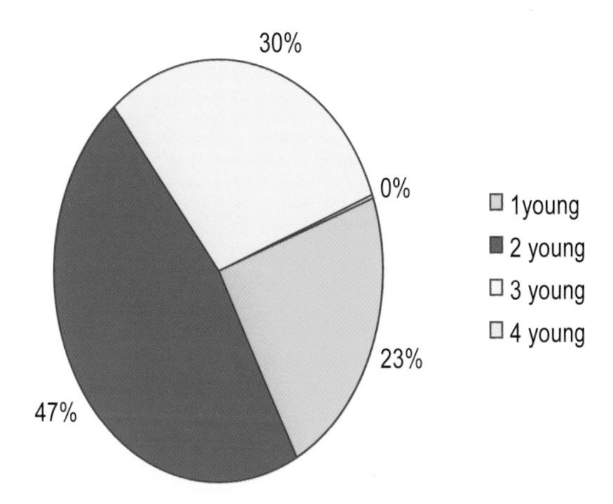

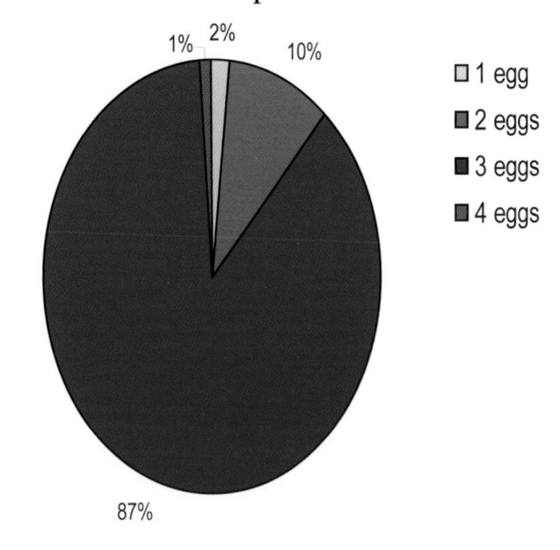

and patterned rather than dark brown like their parents. Their eyes remain distinct amber, quite different from the colour they will become when the birds are mature adults.

The young birds now start to spend more time away from the nest but still remain within a few hundred metres of it, using the eyrie as a meeting place where they receive food. Sometimes, even a week after flying, all the young may come back to the nest and lie down together, just like they did when they were younger. Each now takes fish away to its own favourite perch to eat. Three weeks or so after the youngsters have first flown, their mother leaves on her autumn migration, while the young birds remain and continue to be fed by their father. The most successful broods do no fishing at all of their own, unless their nest is close to the water's edge. The most efficient strategy is to stay near the nest so that they are always ready to be fed as soon as their father flies in with prey.

While they wait for their father, the young are often dispersed within two or three hundred metres of the nest tree, quite often to be found quietly perching low down on fallen trees, stumps or rocks. Perhaps by adopting this strategy, they present less of an opportunity to predators and do not attract the attention of intruding ospreys. The young birds keep a good look out on the horizon for their father and, as soon as he flies in with the next fish, they rush to the eyrie to meet him, while the male leaves immediately to catch a fish for the next chick in the line. At ten weeks of age, the youngsters do not need as much fish as they did during the period when their bones and feathers were actively growing but they appear not to know this, for they are always hungry.

This is the time when the young ospreys are perfecting their flying skills and increasing their body weight in preparation for their migration. When they have been flying freely for about three weeks, they start to soar in the thermals and become more adventurous. Then, after five or six weeks of flight, the oldest chick is ready to migrate and usually leaves at the age of about eleven or twelve weeks, during a spell of good weather. A few days later, the middle

A newly fledged juvenile, September 1991

youngster will follow suit and finally, some days or perhaps a week after that, the youngest chick will depart. Often, sometimes as soon as the following day, the male bird will also leave for his winter quarters, his summer of hard parental duty finally over for another year. The whole family has gone and all its members are now making their separate individual journeys to Africa. For some Scottish young, the first fish they catch for themselves will be in England or France, or even further along on the southward journey.

Some young ospreys die between fledging and migration. It is difficult to know how many, but those suffering food shortages during the summer are most prone to mortality. The two poorly fed chicks hatched at Loch Garten in 1990, for example, died near the nest tree after fledging. We have occasionally found the remains of a previous year's young when first visiting eyries the following spring. I think the level of mortality may be of the order of 5% but a friend in Sweden believes that in his country, it could be much higher. In Sweden, predation by goshawks exacerbates the problem of mortality among young ospreys.

Most ospreys have migrated by the beginning of September and it is rare to find any remaining by the end of the month; occasional ospreys are seen in October and November, but these are vagrants from the Continent. I have seen one in December, an immature reported beside the river at Grudie power station, Ross-shire, on 16 December 1980. We caught it in a very emaciated condition standing on the edge of the river the following day but despite our best efforts it died.

Calls

Ospreys have a range of calls, mainly based on querulous piping and whistles. The male's voice is higher than that of the female and it is quite possible to tell them apart. The main cry is the alarm call, a distinctive loud 'kew-kew-kew-kew', with a varying numbers of calls running together. It can be heard from up to a kilometre away, even by human ears. This call is also used when other raptors or potential predators are sighted and it is particularly explosive when the osprey spots a real threat such as a golden eagle or goshawk.

The female has a squeaky repeated call which she resorts to particularly when she wants the male to go and catch a fish. This is a repetitive squeaky 'quee-quee-quee' and can last for a long time. A similar, but quieter call is emitted by the youngsters when they are hungry and waiting to be fed in the nest. The male bird has a distinctive high-pitched 'pee-pee-pee' which he uses during his display flights.

Both sexes, and also the larger chicks, will use special 'tchuip-tchuip-tchuip' calls when they see an intruding osprey approaching their nest. These calls can be very varied, rapid and high pitched during fighting and they can also be heard during disputes over fishing sites or other territory. When females are disturbed at the nest by humans, some birds give out a grunting call and, in addition, their usual alarm call becomes very aggressive. We have found that individual birds can be either noisier or quieter than is usual and that some have particular individual calls. There are also many softer, piping calls at the nest when the chicks are small.

Logie caught for ringing and fitting of a satellite transmitter using an eagle owl decoy and a dho-gaza net, 2007

6

Migration, ringing and satellite tracking

We may think of ospreys as Scottish birds and look forward to their arrival in the spring as the start of our year, but their casual departure in the autumn, much less of an event for us, is the start of a season elsewhere. The fishermen of Senegal regard ospreys as a talisman – they believe that to have the bones of a fish caught by an osprey on board their boat will ensure a good catch. The Wolof fishermen on that same coast have their own song about this special bird, which translated into English reads:

Osprey, the special one,
Fisherman of the sea,
He does not have nets,
He does not beg for fish.
The fisherman and his boat,
The Osprey and his skills,
There will be no lack of fish.

So we share the osprey with many other nations as it migrates to and from its winter quarters in the south. These great, long distance journeys have fascinated people throughout the ages but it is only in the last 50 years that we have started truly to grasp the intricacies of migration. Now, we have the benefit not only of traditional ringing but also of the latest technological advances to help us understand what the birds experience.

Ringing

In late June and July, we ring as many young ospreys as possible as part of the National Bird Ringing Programme run by the British Trust for Ornithology (BTO). We are fully qualified bird ringers and have the necessary conservation licences obtained from Scottish Natural Heritage which allow us to visit osprey nests. A metal ring inscribed with a unique number and the London address of the British Museum is placed around one leg of each chick. The other leg is fitted with a plastic, coloured ring inscribed with numbers or letters. In the early days, we used a single letter to identify each bird, but now that there are so many more ospreys, we need a combination of two letters or numbers to distinguish one bird from another. The rings are read

Colour rings

vertically and the letters or numbers repeated three times around the ring. Each unique colour ring allows us to track when a bird comes back to breed in Scotland or follow its progress on migration or in Africa. The metal ring is more valuable when a dead or dying bird has been recovered. All ringing details are sent to the BTO database and any report they receive of one of our ringed birds is passed on to us.

Ringing is usually carried out in groups of three as part of the BTO's National Bird Ringing Programme. From left to right: the author, Fiona MacPhie and Jed Andrews.

It is now practice to ring the chicks at a slightly older age than when we first started. We have learnt over the years that this is safer for the bird, and it is also easier for us to work out the sex of each nestling once they are a bit older. Usually working in a team of three, we choose a fine day for ringing, with no rain or high winds. And, just as it has become routine for us, so it has for some of the adult ospreys, who have clearly learned to recognise the signs when we arrive to ring their young.

Some trees can be climbed freestyle, but most need ropes or ladders to reach the nest. Once the climber is up at nest level, he can lift his head over the edge, going gently in order not to panic the young. Usually the alarm calls of the female will already have made them lie still in the nest. The chicks are then carefully lifted out and gently placed in a canvas rucksack, to be lowered quickly to the ground.

In general, young ospreys are a delight to ring, compared with most raptor chicks which can be quite aggressive, even from an early age. There are always exceptions though. About fifteen years ago in Moray, we suffered our first attack while climbing to a nest, and one of those early attacks left the climber needing stitches after the bird's talons raked his face. At this particular group of nests, climbers must now wear a face mask and helmet and there is no reason

Ringing Loch Garten chick at nest A01, July 1983

for the birds to stop attacking: they see it as a successful strategy because after 'driving us off', they find that their chicks are safely back in the nest. The behaviour has been reinforced and will be repeated again in years to come.

Occasionally, while ringing the chicks, we suddenly hear a new set of calls to those of the female. It is the shriller call of the returning male osprey, often carrying a fish (which sometimes adds to the hazard of tree climbing by crashing through the branches towards us). If the male does drop the fish, we make sure we send it up to the nest in the bag, along with the chicks.

Once the chicks have got to ground level and are out of the rucksack, they will usually lie

How heavy are you? – author weighing chick at ringing time

Identifying a colour-ringed individual

Sunday, 23rd April 1995. Nest B15. 1551 hours: The female screaming for food on the eyrie – no rings; male eating fish on T perch, 'blue/black K' colour ring on left leg, BTO ring on right leg. Five minutes eating then dropped skin. Four minutes later flew to nest and successful copulation. Both on nest when I left.

Ringing chicks at a nest near Inverness on a still clear day

quietly together, although the occasional older and bolder one will stand up and peck at our fingers. They will even stand quite meekly on the scales while we check their weight, peering down as if anxious about how much they weigh. Very well fed female chicks can be as heavy as 1800 grams but 1500–1600 grams is the norm while the males weigh anything from 1200 to 1400 grams. We measure the length of the longest feather in the wing and tail as these are useful indicators of age and, along with the weight check, give an idea of the sex of the bird.

We record whether the crop is full, half full or empty, telling us when the bird was last fed. We take note of unusual features like fault bars, distinctive weak marks across the feathers caused by a lack of food for one or more days during the time they were growing. Such food shortages are nearly always due to prolonged periods of heavy rain which make it very hard for the male to fish, and serious marks on the flight feathers can lead to them breaking in two. Missing feathers on the back of the neck, especially on that of the youngest chick, show that there have been food shortages and a degree of sibling aggression. For specialised studies, we may also take a small feather from beneath the wing to be used later for DNA tests and sex determination.

The chicks are then placed gently back inside the bag and pulled up the tree by the climber who has been waiting at the top all this time, in order to get them back to the nest as quickly as

Catching ospreys for research

24th August 1999. Very cold, frost on windows but sunny all day. Collected Joe Hayes at 5.30 a.m., met Alan at fish farm. Six ospreys already there – put out three trap fish (with different breaking strain nylon – fine, 20 lbs, 40 lbs) – caught three ospreys by 7.30 a.m. – first was on second dive at 6.15 a.m. – 'orange/black SO' the male from Loch Insh, and then a Norwegian ringed male – put satellite radios on the two males – then caught a female – not ringed – so ringed it and released. Released both males with radios. Then reset traps but no further luck. About eight males, all adults, came in during the morning as well as a female perched on large tree calling at males, then two of them playing about over the heronry and the male bombing wood pigeons. Left at 8.45 a.m. Really happy with success!

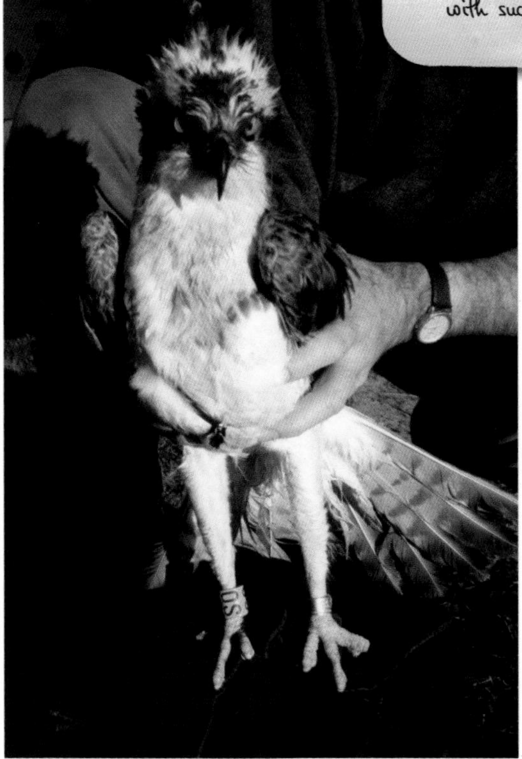

Male SO, August 1999

possible. Once we walk away, the parent birds return to the nest, sometimes almost immediately. Ringing continues throughout early July with late-hatched stragglers still being ringed at the end of the month.

The first ringing of young ospreys in Scotland took place in 1966 and was carried out by Douglas Weir, from whom I took over in the 1970s. Over the decades many friends and other people committed to the return of the osprey have helped with the ringing of the young, and it is now being done elsewhere in Scotland and recently also in England. Over the years we have ringed over 1300 young ospreys in the Scottish Highlands and, to avoid confusion, I remain responsible for the colour ring combinations, given that in Britain as a whole the ringing total now stands at well over 2000 birds.

I have always considered it a privilege to see ospreys at close range during ringing time, and it allows people who have helped look after their local nest to see what their work has achieved. There is a real thrill to be had in seeing the chicks at close quarters and it encourages people to take an even greater interest in their local ospreys, following the lives and journeys of individual birds that they have got to know.

Ringing recoveries

Ospreys migrate in August and September and we usually hear no more about them as individuals. Sometimes, though, we do, as the BTO receives the details of ringed birds from across the world, reporting the findings to the ringer. It is always very exciting to get a glimpse of what happened to the ospreys once they had left Scotland, even though this sometimes involves some sadness at the brevity of their life or the manner of their death.

The colour rings can provide more immediate information on living birds. Sometimes we hear news of colour-ringed individuals early on in their first migration while in England or Europe. In the past we might only have got reports on the colour of a ring, allowing us to say that the bird was probably ringed in Scotland in a particular year. Nowadays, though, bird-watchers have particularly powerful telescopes, they know which ospreys to look out for, and they can give us all the information we need, allowing us to identify the bird in question. If it comes from a well-watched eyrie, we may already know a lot about its personal history.

It is very encouraging that we get a good number of identification reports concerning live ospreys from France, Spain, Portugal and, occasionally, even West Africa. These reports are much more valuable than the traditional recoveries of dead or trapped birds which are sent to the BTO. With colour ring sightings, we obtain an insight into the bird's life and it is likely that the individual may be seen again on other occasions, including at its future breeding site.

Unusual recoveries

I have a copy of the original letter relating to one particular bird. It came from the former Yugoslavia and was addressed simply to: G8121, Inform Brit Museum London, SW7. The short message reads: 'October 20, 1985 Bird was found on Long Island (Dugi Otok) Dalmacija Yugoslavia. Fiting wit eagal or falcon. Left foot 81.' I had ringed it as a chick near Kingussie in Inverness-shire on 2 July 1985. It must have died fighting with an eagle, possibly a white-tailed sea eagle stealing its fish and, most unusually for a Scottish osprey, it had migrated south-east to the Adriatic Sea.

Another report involved a bird we had ringed near Forres on 28 June 1976. Two years later, we received the following letter from Aevar Petersen from the Museum in Iceland (with my comments noted in brackets):

Scheme Ring No.	M 9994 er
Species	Osprey Pandion haliaetus
Age, Sex	pull. 2/2
Status	
Ringing Date	28.6.76
Ringing Place	North Scotland
Ringing Coord.	ca.57½°N. 4°W.
Finding Date	ca.1.5.78
Finding Place	Denmark Strait, between Iceland and Greenland
Finding Coord.	ca.67°N. 25°W.
Finding Details	Landed on boat; later released in Iceland.
Finder	per Reykjavik Museum
Ringer	R.H.Dennis 230578

First seen on 25 April by the crew of the trawler Trausti IS300, *fishing 40 nautical miles west of Latrabjarg in NW West Iceland. The bird boarded the ship 2–3 times in the days to follow. On the night of 27 April, the bird once more boarded the ship and hid under a winch. The bird was captured and put in a box. On 29 April the trawler berthed at Patreksfjordur. A member of the crew contacted Finnur Gudmundsson [a famous Icelandic ornithologist] and it was agreed that the bird should be sent to Reykjavik. This was done on the same day by air. The bird was then kept at the Natural History Museum until 30 April when transferred to a better cage at another institute. There the bird was kept and fed while the appropriate authorities in Britain were contacted. [I remember that Finnur spoke to my old friend, George Waterston, in Edinburgh to arrange for the osprey to be flown to Glasgow, so we could release it the next day in Strathspey. Unfortunately, the Ministry of Agriculture said that this was not possible as it would have to go into quarantine.]*

When action to airlift the bird back to Britain failed, the bird was taken to Hafravatn Lake, just outside Reykjavik, and released. This was on 5 May. Then, nothing was known of the bird until the Museum received a letter from a farmer at Stora-Burfell in northern Iceland, reporting the bird dead: 'It was found in a ditch by the road, possibly shot by wildfowlers.' This was such a sad ending for a young osprey that had missed the route to Scotland while on its way home for the first time.

On beh[alf] of this Museum may we thank you for your co-operation in contributing information about [the re]covery of one of our ringed birds. Here are the ringing details
MUSEUM OF NATUR[AL H]ISTORY
P. O. Box 5320, Reykjavik, Iceland

SCHEME RING NO.		SPECIES	SEX AGE
Brit. Mus. M 9994		Pandion haliaetus	Pull.
	DATE	PLACE	CO-ORDINATES
RINGED	28.6.1976	North Scotland	ca. 57½ N 04 W
RECOVERED	22.5.1978	Stóra-Búrfell, Svínavatnshr., A.-Hún., N.-Iceland.	65.33 N 20.04 W
RECOVERY DETAILS	Found dead in a ditch. Possibly shot by wildfowlers.		
RINGER	R.H. Dennis		
FINDER	Jón Gislason		

But other letters give us good news and one of the most interesting to arrive at the BTO's Ringing Office concerned Osprey G7085, another of the birds I had ringed near Forres, this time in July 1984. It was from Stephen Yip, a project consultant working in the Gambia for CUSO, the Canadian equivalent of our Voluntary Service Overseas (VSO). It was such an interesting story that it made the front page of *BTO News* in winter 1986.

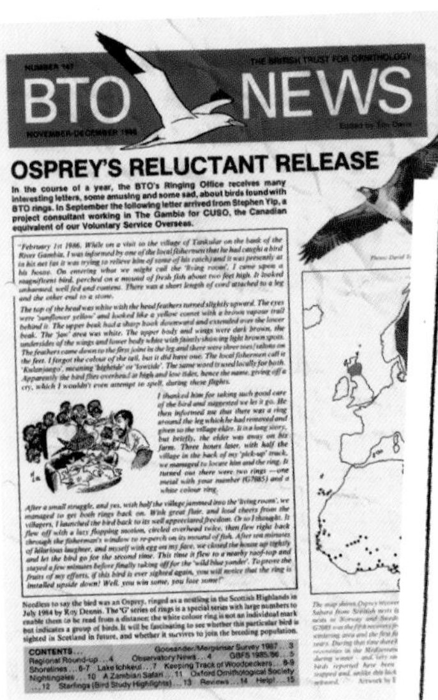

BTO NEWS

OSPREY'S RELUCTANT RELEASE

In the course of a year, the BTO's Ringing Office receives many interesting letters, some amusing and some sad, about birds found with BTO rings. In September the following letter arrived from Stephen Yip, a project consultant working in The Gambia for CUSO, the Canadian equivalent of our Voluntary Service Overseas.

BTO News, *1986*

"February 1st 1986. While on a visit to the village of Tankular on the bank of the River Gambia, I was informed by one of the local fishermen that he had caught a bird in his net (as it was trying to relieve him of some of his catch) and it was presently at his house. On entering what we might call the 'living room', I came upon a magnificent bird, perched on a mound of fresh fish about two feet high. It looked unharmed, well fed and content. There was a short length of cord attached to a leg and the other end to a stone.

The top of the head was white with the head feathers turned slightly upward. The eyes were 'sunflower yellow' and looked like a yellow comet with a brown vapour trail behind it. The upper beak had a sharp hook downward and extended over the lower beak. The 'jaw' area was white. The upper body and wings were dark brown, the undersides of the wings and lower body white with faintly showing light brown spots. The feathers came down to the first joint in the leg and there were three toes/talons on the feet. I forgot the colour of the tail, but it did have one. The local fishermen call it 'Kulanjango', meaning 'hightide' or 'lowtide'. The same word is used locally for both. Apparently the bird flies overhead at high and low tides, hence the name, giving off a cry, which I wouldn't even attempt to spell, during these flights.

I thanked him for taking such good care of the bird and suggested we let it go. He then informed me that there was a ring around the leg which he had removed and given to the village elder. It is a long story, but briefly, the elder was away on his farm. Three hours later, with half the village in the back of my 'pick-up' truck, we managed to locate him and the ring. It turned out there were two rings —one metal with your number (G7085) and a white colour ring.

After a small struggle, and yes, with half the village jammed into the 'living room', we managed to get both rings back on. With great flair, and loud cheers from the villagers, I launched the bird back to its well appreciated freedom. Or so I thought. It flew off with a lazy flopping motion, circled overhead twice, then flew right back through the fisherman's window to re-perch on its mound of fish. After ten minutes of hilarious laughter, and myself with egg on my face, we closed the house up tightly and let the bird go for the second time. This time it flew to a nearby roof-top and stayed a few minutes before finally taking off for the 'wild blue yonder'. To prove the fruits of my efforts, if this bird is ever sighted again, you will notice that the ring is installed upside down! Well, you win some, you lose some!"

Below are just a few accounts of sightings and recoveries of individual birds:

1272542 This was a perfect report of a colour-ringed osprey. Barbara Wadsworth from Swansea wrote saying that she had seen a young osprey at Oxwich, on the Gower Peninsula on 9 September 1990. It was a clear sunny morning and the bird had tried to catch a fish and then perched on a dead tree and preened itself. With the aid of a telescope, she was able to see that it was ringed on both legs. There was a metal ring on the left leg while on the right, a white ring with a large black letter 'C' followed by a dash. She was correct in every detail and this was a young bird hatched near the Cromarty Firth which I had ringed on 10 July.

G8234 Sometimes birds use the same stop-over place year after year while on migration. A male chick ringed near Aviemore on 12 July 1988, with an orange colour ring and a black '4', by then an adult, was seen by Bernard Liegeois at Lac de Jousseau near Millac, Poitiers in France on 26 March and 2 April 1996. He was surprised to see it again during 2–4 September 1997. In the following years it was seen on 21 August 1998, 28 March 1999, 20 March 2000, 27–28 August 2000, 28–27 March 2001, 17–22 March 2002, 17 March 2004, 16 March 2005, 21 March 2006 and 22–25 March 2007. Here was an individual clearly using the same lake and recognising it as a good place to stop over on migration. We never managed to identify his nest site in Scotland. Sadly, on 29 April 2007, he was found long dead on a beach on the Isle of Man.

1306923 Peter Smith saw this colour-ringed osprey on the River Guadiaro, near Sotogrande, Spain on 28 September–2 October 1991. Amazingly, he saw the same bird two years later in the same place on 5–10 April 1993. He reported that on this occasion, it had caught a fish that was about 18 inches long, quite a large catch for an osprey.

1272574 Multiple sightings were reported in the Gambia of a young bird ringed by Steve Cooper in Perthshire on 26 July 1990. The colour ring was read by visiting birdwatchers at Kotu Stream between 17 November 1990 and 7 March 1991. It was wintering there again during 9–18 December 1991 and seen for a third winter from 4 December 1992 to 20 February 1993.

1311944 A chick ringed by Keith Brockie at Loch of the Lowes on 3 July 1997 was seen on 4 September by Roy Frost at Welbeck Lakes in Nottinghamshire. Twenty-four days later the bird was found with a broken wing in the sea at Malika, north of Dakar in Senegal. A detailed letter and photographs told us of the sad fate of this youngster.

1323469 A chick I had ringed near Dingwall on 17 July 1994 was identified three years later, on 19 April 1997, when the colour ring was read by a Mr Christophers at St Columb Major, as it flew north through Cornwall on its journey home as a new, potential breeder. Sadly, on 8 May it was found dead with a broken wing below the Mount Eagle television mast on the Black Isle, just over ten kilometres from its original nest.

1085622 This bird, ringed on 8 July 1980 near Forres, was killed by an agricultural worker on 17 November 1990 near Monchon in Guinea. The report came from the forester at Conakry. The bird was ten years old when it met its end and some 5362 kilometres from its natal nest.

1348922 This was a chick collected in Easter Ross and translocated to Rutland Water, where it was released on 1 August 1997 and from where it then migrated on 4 September. The following year, on 4 August 1998 it was intentionally killed near Conakry in Guinea.

1348992 A very unusual record of a chick ringed at Nethybridge on 5 July 1998: This was a large female in excellent condition weighing 1650 grams with a wing length of 350 millimetres and she must have fledged in mid-July. On 30 August 1998, the bird was found alive with a broken wing on a road near Wroclaw in Poland but she later died. The bird's location, at a distance 1505 kilometres to the east, was a very unusual finding.

1272512 Howard Orridge of Melton Mowbray wrote that on 26 July 1991, he was walking along the River Findhorn and came across a 'buzzard' hanging from a branch by a fishing line. This had become entangled around its left wing and leg and had then caught on a branch when the bird attempted to perch. The osprey was fitted with a yellow ring on its left leg with three black 'U's on it and a BTO ring on the other leg. It had been dead for some time as several insects emerged when he retrieved the body. I had ringed it as a chick near Elgin on 6 July 1989 and so it had met this sorry end on its first return to the Highlands at two years of age.

G7055 An unusual spring record of an osprey ringed in Perthshire on 3 July 1984. It was discovered recently drowned having being caught in a fishing net on 22 April 1988 in Sweden. The attached police report said that a Mr Johansson had taken up his fishing net in Lake Viksjon and discovered the dead osprey inside. At four years of age, this bird could easily have been breeding in western Sweden.

A Norwegian bird

4th July 1998. Nest A06. Female now on nest feeding young, but mainly in tree beside nest. Two big young on nest, smaller low down. Very good look at female with telescope, she has a Norwegian metal clip ring on right leg and I can see the last three numbers are ...350. She has three black spots in each yellow eye, two to the front of the pupil and one below.

The *Migration Atlas*

In 2002 the BTO published the new *Migration Atlas*, which analyses and presents information on the migrations of the birds of the British Isles. My chapter on ospreys was based on analyses of the recovery data up to that time. Birds ringed in Scotland had been found in the following countries: England (23), Wales, Ireland (3), The Netherlands, Belgium (2), France (11), Poland, Croatia, Spain (6), Portugal (4), Morocco (2), Algeria, Mauritania (2), Senegal (4), the Gambia (5), Guinea (2), and single birds in Guinea-Bissau and Benin. There were also single recoveries in the summer from Norway, Sweden, the Faroes and Iceland.

Many of the recoveries were of birds under one year of age and most were less than four years old. The oldest birds recorded were: 10 years (2), 11 years (1), 13 years (1), 14 years (2), 15 years (1), 16 years (2) and the oldest recovery of all was aged 25 years and 7 months. I myself ringed this bird in north Perthshire on 20 July 1974 and it met its death by being electrocuted on overhead cables near Agadir, Morocco, on 15 March 2000 on its twenty-fifth northward migration. Adding together just its direct journeys from Scotland to Senegal, it had travelled a total migration distance over the years of 250,000 kilometres. There is a record of a ringed bird of 32 years of age from North America.

Many more ospreys have been ringed in Sweden than in other countries; 18,214 up to the year 2001, of which 2120 have been recovered. All winter recoveries recorded were from West Africa except for three notable exceptions in East Africa. Of all recoveries 43% were killed by hunters, 10.4% died in fishing nets, 7% were electrocuted and for the rest the cause is unknown.

Ospreys ringed in other countries are sometimes found in Britain. The natal origin of these birds was: Sweden (15), Finland (3) and Norway (1) and, subsequently, a German bird summered in central Wales. Some of these ospreys, of both sexes, from Sweden and Norway have joined the Scottish breeding population. Most of them were probably birds on spring migration that had been driven across the North Sea by bad weather. Being exhausted, they remained in Scotland and subsequently nested there.

Autumn migration

We know that adult females and non-breeding birds migrate southwards throughout August while the young and their fathers leave later, from mid-August onwards, with all birds departed by the end of September. Travelling independently and not as a family, the birds take various routes. Ospreys migrate on a broad front and unlike some other species, do not congregate at places like Gibraltar. Active day migration is at a relatively low altitude of about 500 feet. The general pattern of recoveries and sightings shows that Scottish ospreys fly south in autumn through the United Kingdom into France, Spain and Portugal. The first southern recovery was on 14 August in Hampshire. After crossing to North Africa, ospreys may cross the Atlas Mountains and then, instead of following the coast southwards, head out over the vast deserts of the Sahara to Mauritania, Senegal and the Gambia. Some may move inland to Mali and a few travel even further south to Guinea. The first birds reach their wintering grounds in West Africa in late September. On migration and while away, they meet many other ospreys from the larger breeding populations in mainland Europe.

Increasing numbers of reports suggest that some of our Scottish youngsters move south through Ireland and can make successful journeys direct from there to northern Spain. But we have also had two recoveries from fishing boats, one 200 kilometres south-west of Brittany and the other on the Great Sole Bank, 230 kilometres west of the Isles of Scilly. So it is to be expected that some of the young ospreys die at sea when weather conditions are bad. It is a fact that ospreys nesting in Scotland are more likely to undertake long sea-crossings than are their continental cousins and in so doing, are likely to suffer increased mortality in bad weather. Surprisingly, some young head off eastwards on unusual journeys to Poland and Croatia.

Winter quarters

In 1977, when we were both working for the RSPB, I went to the Gambia with Hugh Miles to make a film about ospreys in winter. We found that many ospreys were living in the mangrove swamps and along the Gambia River and its estuary. These large river estuaries and wetland areas of West Africa are the preferred habitat for over-wintering ospreys. The rivers meander through mangrove swamps as they approach the sea and, along with the coastal bays and beaches, provide an easy living for the ospreys during this important period in their annual calendar.

There are many other fish-eating birds, such as pelicans, herons, terns, cormorants and skimmers, to be found along the coast of Gambia. Although ospreys may hunt from tree perches alongside the water channels in the mangrove swamps, most were observed flying out to fish in the open seas of the Atlantic Ocean. On many occasions, we watched them fishing close to the coast, especially around the estuaries where there were large numbers of terns and gulls and bottle-nosed dolphins, just like at home in the Moray Firth.

There were many fishermen working from open boats along the coast, and we could see that stocks were plentiful. The ospreys favoured various species of mullet and large sardine; at least ten different types featured in their diet. At other times, ospreys flew far out to sea and out of sight but would return later, each with a flying fish, its 'wings' and long 'tail' trailing out behind. Once or twice we saw an osprey carrying a flying fish the wrong way round, dragging it along like a parachute. The poor bird would then land utterly exhausted on the first sandy beach it encountered to rest and eat its catch.

The long sandy beaches along this coast are fringed with palm trees and scrub, the high tide lines dotted with great logs and dead trees that have been swept down the huge rivers of Africa. These were highly favoured resting areas for the ospreys between fishing trips and where they spent a lot of their time during the scorching heat of the West African day. In some favoured places up to 25 ospreys might be seen resting together in a loose flock.

Most birds returned to the mangrove swamps, favoured both during the day and at night. The ospreys chose the dead tops of taller trees within the swamps as their favourite roost sites. When travelling by small boat through the channels amongst the mangroves, we were quickly able to identify favourite roosts, but we looked for colour rings without success, and while we started to recognise individual birds, those we saw were invariably adults. It also became evident that individual birds protected their favourite perch from other ospreys of both sexes. The birds would regularly go out to sea to fish and return to feed on their own favoured tree. Ringing recoveries and satellite tracking have now shown that adult ospreys often winter in exactly the same location year after year, having first discovered a good place when they were juveniles. Not only do they breed every year in a home tree in Scotland but are similarly faithful to favourite trees in Africa. In the winter, though, adults lead a solitary life; they do not even necessarily winter in the same area as their partner.

In Africa ospreys often find themselves pursued by other birds once they have successfully caught a fish. These are not just gulls as in Scotland, but also black kites and Caspian terns, while further inland, they may encounter fish eagles as well. These birds press home their attacks very strenuously but we observed that the ospreys were nearly always successful in keeping their catch. I was particularly struck by the widely varying fishing ability between individual ospreys. During one period of study, I had been watching an osprey which had spent over an hour trying to catch a fish in the estuary of a small river near Banjul; it made 32 dives without a single success. During this time, another osprey flew out from the mangrove swamp at a height of about 200 feet, performed a shallow dive straight into the water and came out with two fish, one of which it dropped. Within two minutes it was heading back to its roost to feed while the other bird fished on.

The ospreys' life along the tropical shores of West Africa appears idyllic. One sometimes wonders why they bother to return to cold, snowy Scotland but the call of home is very strong. Most juveniles are believed to spend their first summer in Africa south of the Sahara, and our Guinea recovery supports that view. In summer 1997, though, a Scottish colour-ringed osprey was seen at Rutland Water, having returned when it was just one year old. In 2003 a translocated female from Rutland Water that had wintered in Portugal also returned to England at the same age. Most recoveries, however, show that two years is the normal age of return to Europe.

Colour ring sightings and ospreys caught by Spanish bird ringers have started to reveal that greater numbers of the birds are now wintering in the Iberian Peninsula instead of West Africa. There are probably several hundred birds that do this, with the salt marshes around Odiel being a favoured site: birds from Scotland, Germany, France and Sweden have been recorded wintering there. During one winter, I joined the National Park team in this region where Jose Sayago ushered me into a hide beside one of his osprey live traps. When I later removed the captured osprey for study, I found it had been ringed in the Scottish Borders. More males (70% more) than females are recorded over-wintering here but there are about equal numbers of young and older birds. This appears to be a growing phenomenon and may be due to warmer winters caused by climate change. Even further north in France, one and sometimes two ospreys have regularly wintered over the last ten years in southern Brittany south of the town of Brest. They arrive in late August and do not leave until March. If the winter weather continues to warm, it would not surprise me if the occasional osprey winters in Devon or Cornwall in the near future. These shorter migrations are probably beneficial to ospreys and may result in higher survival and greater longevity.

Spring migration

The adult ospreys begin to move north again in March and now there is no time to linger. Their ancient instincts drive them on and the spring migration is much faster than the one they made in autumn. The birds are driven by the need to get back to claim their old nest and to pair up again with their previous year's mate, before another osprey tries to take their place. The birds use the same return routes across the Sahara desert and head out of Morocco into Spain. While most cross the Pyrenees and fly through France in order to reach England, there is no doubt that many choose a shorter journey, heading out of Brittany and from there across to Devon or Cornwall, or maybe even flying from northern Spain across the Bay of Biscay.

Once they arrive in southern England, they do not rest but continue to press northwards to reach their breeding sites as quickly as possible. But this eagerness to get back home, sometimes by taking short cuts over the ocean, can result in problems during bad weather and may sometimes even prove fatal. Two of our ringed birds missed landfall in Scotland altogether and were reported in the Faroes and off Iceland. This type of mishap is undoubtedly one of the causes of loss of adult ospreys and the following account reveals how one such bird miraculously recovered from an almost certain death.

An adult male breeding in Strathspey departed on 30 August 2000. After 22 days he had reached his wintering quarters at Sine Saloum in Senegal, having travelled a distance of 5317 kilometres. He started his return journey in mid-March and by the 27th of that month, had travelled 2562 kilometres north to southern Spain. The next report of him was out in the open North Atlantic Ocean, 160 kilometres west-north-west of St Kilda, which was where he was reported – thanks to his radio transmitter – at 0300 on 6 April. He had run into bad weather and this had driven him out past Ireland as he was heading northwards. Two days later, on 8 April, he was near the Monach islands off South Uist at 1400 hours while four hours later, he reached the Point of Sleet in southern Skye. After such an incredible and dangerous journey over the ocean he must have been very weak, because it took him until 28 April to return to Strathspey. He had flown just 140 kilometres in 20 days, only to find that his mate had not waited but had

First time using eagle owl to catch breeding female

26th July 1998. Caught breeding female at nest A11 near Carrbridge using an eagle owl decoy. Owl kept jumping off the perch when osprey dived so put out a stuffed owl as well and quickly caught the adult female osprey in the net. She was caught by her talons. Found out she was already ringed; number '1306910' on left leg and dark green-white colour ring J on right leg. Her weight was 1810 g; wing length 524 mm and tail 232 mm; she was in excellent condition and really plump ready for migration. Attached satellite radio '21197'. Released her about a kilometre away from my Land Rover and she flew straight back to the nest and landed on it with the young. Found that I had ringed her as a chick 10th July 1991 in Ross-shire at nest G07.

2nd November 1998.

Extremadura, Spain. Driving along lake side when I saw an osprey flying along the lake chased by jackdaws. Stopped by a small harbour, some houses and a restaurant to watch the osprey hunting about 150 feet above the water in bright sun. Suddenly saw the aerial of the satellite radio as she flew past me and then she flew away to the south of the reservoir. Location: 39° 44'north; 6° 27'west.

paired with a three-year-old osprey, laid three eggs and started to incubate! But he fought off the intruder and then helped to raise a single chick. The following year, he made it home on time.

Bad weather with easterly winds blowing offshore in northern Spain, the Bay of Biscay or southern England and Ireland, especially during the first two weeks of April, can cause the loss of breeding adults on their spring migration. This was the case in April 2005, when more 'regularly breeding ospreys' than usual failed to return to their nests in Scotland or were very late in arriving. The year 2007 was even worse. It is also possible that a very small number of Scottish ospreys may unusually take a more easterly flight route north and become assimilated in the Norwegian and Swedish populations. We have one spring recovery from Sweden, which suggests that this might have happened. So far, none of our colour-ringed birds have been identified as breeding in mainland Europe.

Satellite transmitters

Traditional radio tracking of birds has been used for several decades but, during the 1990s, technological advances allowed the manufacture of very small radios which could communicate with satellites circling the earth. These lightweight transmitters, weighing just 30 grams, had been developed in North America and fitted to migratory ospreys in the United States and Sweden, revealing incredible detail about their migrations.

The radios transmit signals at set times and these are then picked up by satellites. The French company, CLS, based in Toulouse, established the Argos system that collects and analyses the data. This can then be downloaded to a computer once a contract has been established with CLS. The costs are high but the information received is often of very high quality. The batteries

Catching ospreys for satellite tracking – the female was caught using an eagle owl decoy

Female from nest site A11 fitted with satellite transmitter

in the transmitters last for about 800 hours in the standard models that we use, allowing one hundred days or more of transmissions. These can be pre-programmed by the manufacturer, in our case Microwave Telemetry in Columbia, to transmit at different intervals of between one and ten days.

During the time of our Rutland Water project, one of the partners in the group, Anglia Water plc, agreed to fund a satellite-tracking programme. We wanted to discover whether the translocated young ospreys were reaching their correct winter quarters in Africa as well as understand more about their migration routes, *en route* stopping places and wintering locations. (This information is presented in Chapter 10.) It was also agreed that we would examine the migrations of the donor population in Scotland, preferentially concentrating on the parents and siblings of the translocated youngsters.

Female ospreys

The first osprey that I caught for our satellite tracking research was an adult female nesting near Carrbridge in Strathspey. I had originally ringed her as a chick in Easter Ross in 1991 and she had been breeding successfully since 1996. In 1999 she had two young and when these were well grown, I decided to fit her with a radio. After catching her in a net near her nest on 17 July, having first obtained the necessary licences, I weighed and measured her, then carefully fitted the satellite radio onto her back like a tiny rucksack. The four Teflon ribbons used to secure it were sewn with pure cotton underneath her breast band. The cotton would slowly rot and allow the radio to fall off after a couple of years, when the battery was exhausted.

I released her about one kilometre from her nest. As she flew off she shook herself and made straight for her nest tree, landing safely back beside her young. I saw her on many subsequent occasions until suddenly, one day in August, she had gone. I knew that she had still been there on 6 August and the next information came in on 25 August, when she was at a big reservoir in Extremadura, south-west of Madrid. She was still there on 28–30 August and well into September, while we continued to anticipate news of her departure. But, eventually, we realised that this was where she was going to stay throughout the winter.

In November I went to Spain and drove to the area from where the signals came. I saw just one osprey at this huge reservoir, Embalse de Alcantra, and sure enough it was 'my' bird, being chased by jackdaws as she searched for fish. Through my telescope I could see the tiny radio aerial sticking up from her back. She caught a fish and flew off into the cork oak forests to eat her catch. It was such an exciting moment to find an osprey that I knew so well from back home. Her favourite roost site was a huge nest in an isolated tree that had been built by a white stork and, thanks to satellite tracking, I came to know so much more about her than I would ever have expected.

She was back at her nest in Strathspey on 6 April and reared another two young. She then departed again on 10 August and, after a six-day stop-over in the Scottish Borders, migrated south via England, the Cherbourg Peninsula, the western Pyrenees and so to the same reservoir in Spain, having flown 2800 kilometres in 17 days. She returned home again the following spring, arriving on 4 April 2001 after a 15-day journey. We then removed the radio, but not before she had demonstrated the osprey's fidelity to both its summer and winter quarters and the similarity of its annual migrations.

We quickly realised how much the use of these satellite radios was teaching us, and we decided to catch and fit some more birds with these devices.

Two more breeding females were tracked from their nesting sites in the north of Scotland. 'S01' was a three-year-old breeding in Strathspey for the first time. She departed on *c.*26 August and flew steadily southwards to south-west Spain, where she stopped over for a few days before travelling fast through Africa, to reach St Louis in Senegal on 27 September. Her migration took her over 5000 kilometres in just 30 days. She set off north again on 28 March and a series of 'dog-legs' took her right across the Sahara desert. She arrived back at her nest site on 22 April, having covered 5157 kilometres in 25 days.

The other female, 'S07', was breeding in Nairnshire and was of unknown age. She migrated sometime in late August and flew steadily south through France and Spain, then taking a more interior route through Africa across the centre of the Sahara desert, before turning westwards in to Mauritania. She then finally wintered on an island off the coast of Guinea-Bissau, arriving there on 29 September after flying at least 5741 kilometres in 27 days.

S07, 5741 km in 27 days

Male ospreys

I fitted satellite radios to three different males which I caught at Rothiemurchus fish farm. Two were local breeding males and the third was, interestingly, a young osprey that had been hatched near Oslo in Norway two summers previously. 'S10' was caught in July 1999 and was breeding in Badenoch; he was eleven years old and had been ringed as a chick nearby in Strathspey. This was an experienced male that left on 13 September and flew south in a leisurely manner to winter in Mauritania. He rested for three days in Wiltshire and had a six-day stop-over in south-west Spain, before setting off at dusk to fly 600 kilometres, straight across the sea to Morocco, during the night. In the spring of 2000, he set off north on 22 March and flew the 4547 kilometres back to his nest in 18 days. He set off south again on 27 August and reached exactly the same winter quarters on 29 September. In spring 2001 he set off for the north on *c.*24 March, but he must have got lost in bad weather as he did not reach his nest until 26 April, nearly a month late, only to find it occupied by a pair of interlopers. He regained his nest but did not breed. The following spring he was back on 30 March to make certain of his rightful claim to the nest.

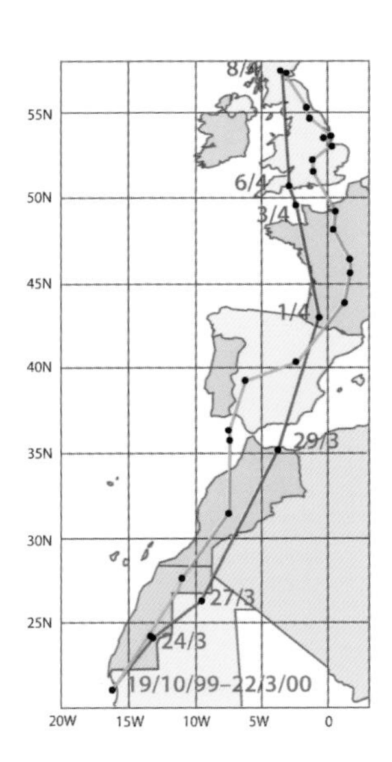

Migration of S10, 1999–2000 Autumn (orange), Spring (red) – 4547 km in 18 days

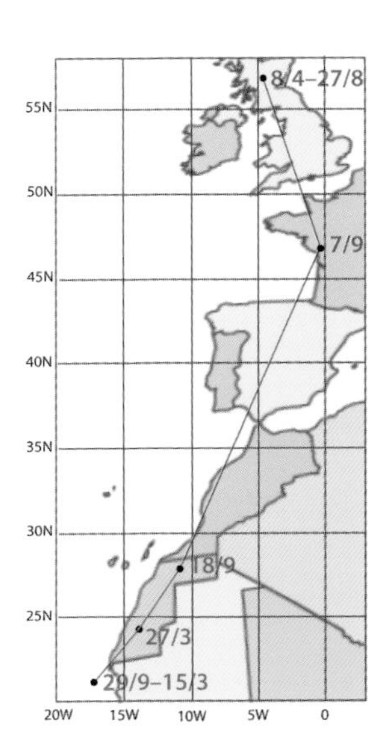

Migration of S10, 2000–2001

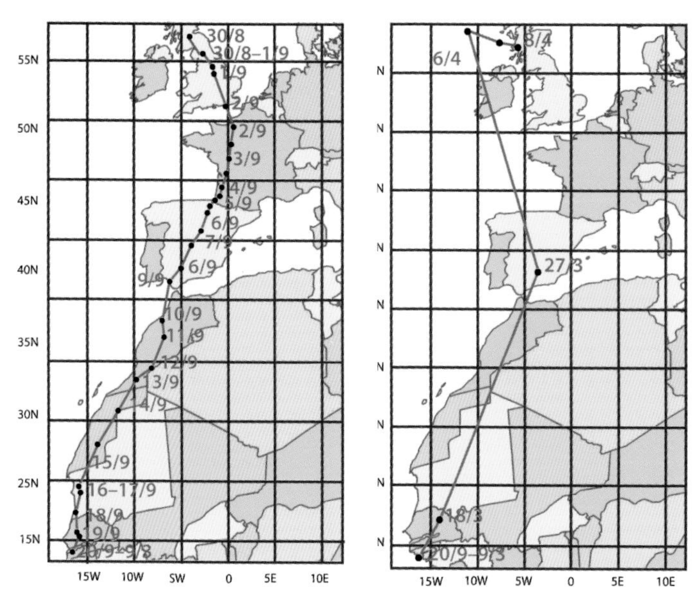

Far left: Autumn migration of S18, 2000 – 5317 km in 22 days

Left: Spring migration of S18, 2001; 2562 km in 9 days, then lost at sea. By 8 April found his way to Skye, before eventually returning to Strathspey on 28 April – a distance of 140 km in 20 days

'S18' was a breeding male, mated to 'S06', nesting near Carrbridge in 2000. He set off south on 30 August and, in just 22 days, covered 5357 kilometres to reach his wintering quarters in Sine Saloum in Senegal. He started north on 9–18 March and then got lost in the Atlantic, consequently returning home rather late. The young Norwegian male left Strathspey on 3 September and travelled very slowly south, having a 17-day stop-over in the Lake District and then spending three days around Le Havre in France before he finally reached Guinea-Bissau on 17 October, having flown a distance of 5700 kilometres. He returned to Strathspey the following summer and left Scotland on 8 September 2000 to arrive again at exactly the same wintering site on 10 October.

Juvenile ospreys

Seven young ospreys have been tracked from Scotland and their journeys have proved very revealing. The first, in 1999, was 'S08', a young male from a nest in Nairnshire. He set off on migration sometime after 3 September and on 10 September was passing over Kintyre, heading towards Ireland. The next transmission was from the border between southern Spain and Portugal, but after reaching Lisbon, the radio stopped working. I was certain that this bird migrated through Ireland directly to Spain but the radio was only sending signals every seven days and so we missed filling in the details of the journey.

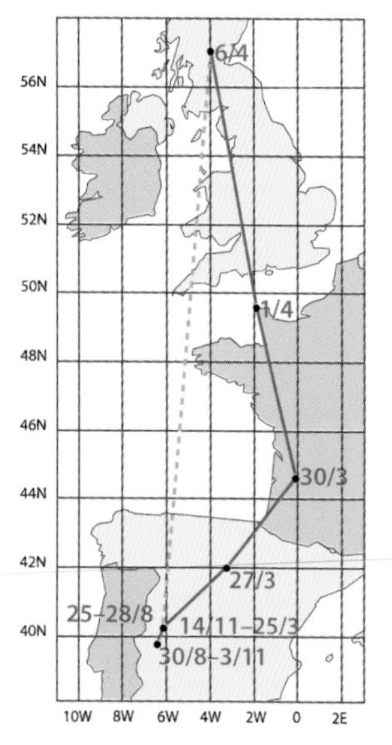

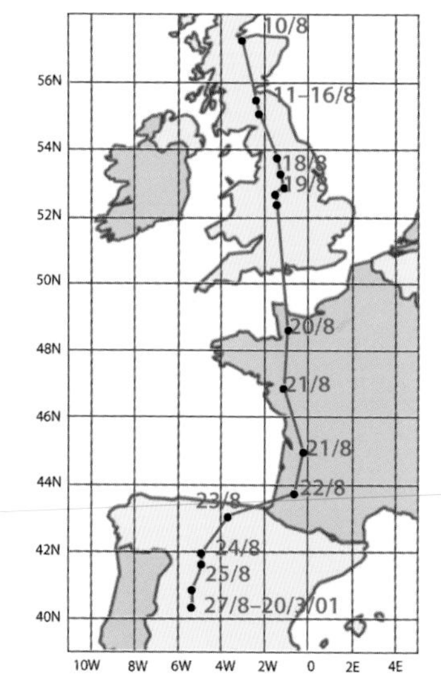

Far left: Migration of S06 – mated to S18 – Autumn 1999 (orange) and Spring 2000 (red)

Left: Autumn migration of S06, 2000

In 2000 another juvenile male, 'S11', from a nest near Carrbridge, showed us that such journeys were, in fact, possible. He departed on 9 September and was in south-west Scotland by the tenth. He then travelled down the east side of Ireland on the twelfth. He left the south coast of Ireland at 1300 that day and 15 hours later he was 500 km from land, half way across the ocean. He flew non-stop on 13 September and finally reached the Spanish coast at midnight after a continuous flight from Ireland of 1008 kilometres in less than 36 hours. This was a truly remarkable journey for a young osprey. He flew on to central Portugal where he rested for two weeks before leaving from the south of the country on 30 September. He flew straight out over 600 kilometres of sea to reach north-west Morocco before making a straightforward migration down across the deserts to Sine Saloum in Senegal before wandering further south along the coast to the border with Guinea-Bissau.

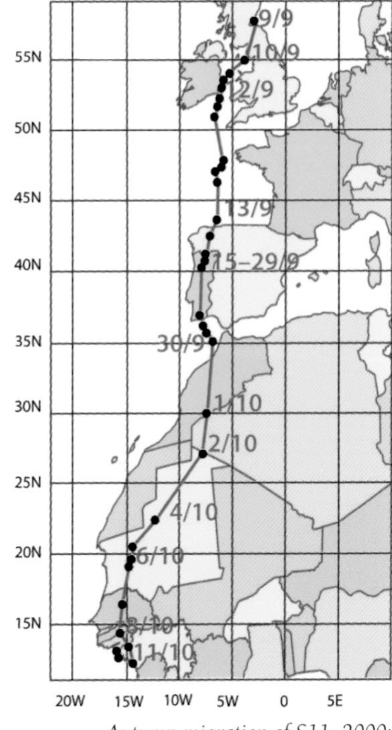

Autumn migration of S11, 2000: 1008 km in 36 hours

Another young male left Strathspey that autumn at the end of August, reaching Surrey on 5 September and then, very surprisingly, remaining near Aachen on the border of France and Germany from 14 September to 17 November. In late November he re-started his migration and skirted the western Alps on the 28th. His next position was in the central Sahara on 10 December and on the 20th, when the batteries of the radio ran out, there were still 500 kilometres to go in order to reach Malia.

In 2001 we tracked two youngsters but one had a radio failure on 24 August after it reached Evesham. The other was in some ways similar to the German report featured above. The bird left its nesting site on the Dornoch Firth on 3 September, roosted near Inverness on the fourth and reached Dundee on the fifth before very bad weather forced it north to the north-east coast of Aberdeenshire. On 8 September it flew south-east across the North Sea, arriving in the Friesian Islands late in the afternoon. Continuing in a south-easterly direction it made a temporary home in a landscape of lakes and rivers south-east of Hanover. Finally, on 17 October, it headed south-west through France, crossing the Mediterranean from southern France and heading into Algeria. By 5 November it was in the deserts in central Algeria, and by 15 November had arrived in eastern Mali, about 40 kilometres north of the River Niger, a good wintering site for ospreys. This journey of nearly 5000 kilometres had been accomplished in 36 days of travelling, with a 36-day rest period in northern Germany.

To try to test the effects of parentage, I tracked two offspring of a Norwegian ringed female breeding in Nairnshire. The first was a juvenile male osprey, the younger of a brood of two. He departed on 2 September 2003 but only moved about 20 kilometres to the River Spey where he lived until 1 October. A surprisingly short distance and an extended stop-over! By the evening of 1 October, he was roosting near Blairgowrie in Angus, some 95 kilometres to the south and the next day, he flew 150 kilometres south to Seaham, Durham. At 0855 on 4 October, he was south of Dieppe and then made a fast journey through France and Spain, being west of Toulouse by the late afternoon of the fifth, with a timed four hour journey of 150 kilometres at a speed of 37.5 kilometres per hour. On the night of the sixth he was in eastern Spain near Albacete and the following evening he roosted just north-west of Malaga. He crossed the Straits to Africa on the eighth and spent that night near Quezzan in north Morocco. By the evening of the 12th, he was heading across the Sahara desert in Mauritania and the next evening, on the 13th, he was further south near Fderik. On the early morning of the 16th, he was still travelling through the deserts 100 kilometres east of Nouakchott. By 18 October, he had reached the Senegal River near Rosso in southern Mauritania, where he spent the winter.

A year later I tracked a juvenile female from the same parents – she was designated 'SSK'. She departed during the afternoon of 12 September, once morning fog had cleared to bright sunshine with the advent of a high pressure system over the whole of the UK. She travelled south over the Cairngorms and roosted overnight near Braemar. On the next day she set off in a south-westerly direction in clear skies and little wind, passing the Forth Road Bridge at 1300 and crossing the Kirkcudbrightshire coast at 1800 to reach the north end of the Isle of Man less than an hour later. She had flown about 300 kilometres during that day alone.

High pressure and clear sunny skies were perfect for migration and on the morning of 14 September, she left the Isle of Man and reached Holy Isle, Anglesey, at 1022. The Dyfi estuary, north of Aberystwyth, was passed at 1406. At 0404 on the following day, she was roosting near Lynmouth in north Devon. Excellent weather continued with clear skies and light winds. The bird was in Bideford Bay at 0745 and the next signal (the radio was programmed 8 hours on and 8 hours off) was at 17.17 p.m., when she was about ten kilometres from the most south-westerly point of Brittany.

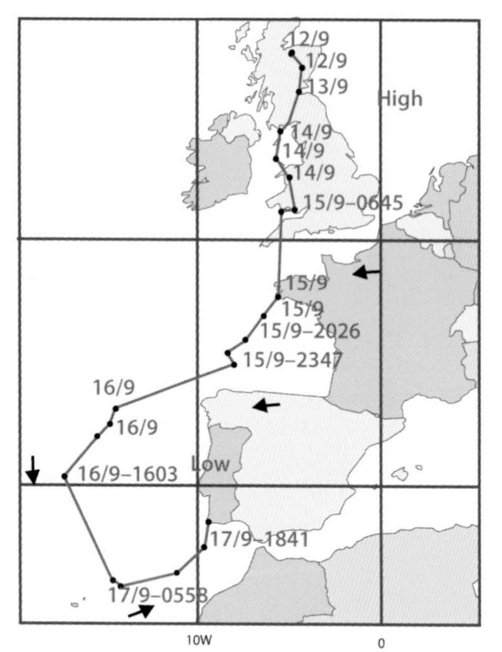

The young SSK's migratory journey, 2003: 3000 km covered in 60 hours

The bird set off across the Bay of Biscay and by midnight was 200 kilometres north of Ribadeo on the north coast of Spain. She had flown nearly 600 kilometres since starting the day's journey in north Devon. By this time, the young osprey was travelling over the sea at night and had entered a vigorous low pressure system centred to the west of Portugal. She was being blown out into the Atlantic Ocean by strong easterly winds under thick cloud.

On 16 September she was 400 kilometres west of north-western Spain. She missed reaching the north Spanish coast and during that day, migrated downwind in strong northerly winds, at 1603 about 600 kilometres west of Lisbon. Next morning, after flying for another night over the ocean, she was 250 kilometres to the southwest of Portugal, but by now the winds were westerly. Late in the evening, a very tired osprey made landfall on the south-west coast of Portugal near Odemira. She had flown non-stop through the day and night for about 60 hours and travelled about 3000 kilometres over the sea.

This young osprey survived an incredibly difficult migration. A day later she moved a little further north and subsequent signals from large reservoirs and rivers south of Lisbon showed that she was fishing and resting. She decided to winter in southern Portugal and data through to 15 March showed that she moved around various reservoirs and rivers in the southern half of Portugal. In March the radio's batteries were exhausted and transmissions ceased, but it was very encouraging to have learned that this youngster survived such a dangerous journey.

The information from the satellite transmitters has already given us much new information. The adults are experienced travellers who know the best routes and stop-over sites for feeding *en route*. They generally head east of south out of Scotland in autumn and then continue across the narrower parts of the English Channel into France on their journey south. Most winter in West Africa as far south as Guinea but there is the new trend of increasing numbers wintering in Spain or Portugal. On the return migration, when hurrying northwards to claim their nest sites, they sometimes cut the corners across wide stretches of ocean heading for southwest England and it is likely that this sometimes results in birds being lost at sea in the North Atlantic. We also now know that ospreys readily cross broad stretches of water to reach North Africa and do not need to go to the narrows near Gibraltar. They sometimes even fly at night but generally travel by day.

Satellite studies on Swedish ospreys showed that these birds spent an average of 45 days travelling an average distance of 6742 kilometres to West Africa. It was thought that the birds were travelling between 8 a.m. and 5 p.m. and migrating between 250 and 360 kilometres per day. It has been shown that adults have a much greater ability to adjust their flight direction than do juveniles when they are drifted by strong winds. One of the Finnish adult males, which breeds just north of the Arctic Circle in Lapland, probably makes the longest migration of all ospreys, some 12,564 kilometres to the southern tip of Africa.

The juveniles, on the other hand, show much more variation. They are making their first journeys relying only on their genetic information. They tend to have a more south-westerly heading when they set off from Scotland. This sometimes results in long migrations over the sea, including direct journeys from Ireland to Spain, which is a surprisingly long journey for a young raptor that has only been flying for about six weeks. On the other hand, two birds moved to the south-east and rested in Germany before continuing south. There is no doubt that there are losses on the first autumn migration and the fact that we are an island off the continental coast means that more of our native birds are likely to suffer from weather-related deaths than those migrating from Scandinavia and Germany. Ospreys from Sweden tend to move south-westwards to reach Spain and West Africa. There is a strong genetic input into the increasing Scottish population from ospreys of Scandinavian origin and this may be the reason that there seems to be a surprisingly high chance of some of our youngsters heading down over the Atlantic Ocean instead of taking the easy route across the English Channel. In the long term, evolution should sort out this problem. We also know that some juveniles move about in the winter, suggesting that they are searching for the best possible wintering sites.

There is much more to learn about the behaviour of juveniles in winter and during their first summer, and about the spring migration for birds of all ages. We need to know more about the importance and conservation status of stop-over sites, the extent and frequency of voluntary and involuntary sea crossings, the effects of man-made hazards, especially overhead power lines on estuaries and fresh waters and the location and conservation status of the most important wintering sites in Africa. We, in Europe, also need to establish strong partnerships with conservationists and protected areas in Africa, where our birds spend half their lives.

Postscript

In the summer of 2005, David Anderson of the Forestry Commission found a new osprey nest near Aberfoyle and saw that the female was carrying a satellite transmitter. At a later date, he saw the colour ring and we identified it as this very special bird – osprey 'SSK' 2002. Nothing more had been heard of her until David's exciting news. So this amazing individual not only survived her dangerous first migration but ultimately returned to breed in Scotland, giving us a very happy ending to the story. She returned again in 2006 and 2007, although she moved to a new nest.

7
Ospreys as individuals

When we look at a species from a conservation point of view, we think of numbers of pairs, their distribution and the number of young reared annually. How do they perform as individual members of their species? Which ones return to breed, how long do they live, how successful are they as parents and are their offspring successful in turn? All important questions, but as I have followed the fortunes of ospreys, I have come to realise that they have highly individual characteristics and behave in highly individual ways. Generally speaking, adult ospreys return to the same nest each spring and usually pair up with the same mate, but we can only know that by getting to know individual birds, identifiable not only from their coloured rings but also from their plumage.

As early as the 1970s I realised that I could recognise individual breeding ospreys from their markings, and sometimes even from specific behaviour. Looking at them closely, year to year, I began to see individual characteristics in each bird. I made a small sketch of each of them at my regularly monitored nests and checked them in subsequent seasons. The markings

> Pairs moving their nests
>
> 21st April 2000. North Sutherland. Colin Crooke and I to Angus's house; then to old nest and to last year's nest both empty. Colin climbed up tree at the old nest and cut away some growing branches obscuring the nest – then male osprey above us and also buzzard, peregrine and sparrowhawk. Then we went round to the other woods where the shepherd had seen a bird going recently – found nest on Scots pine in clearing with a pair of eggs – female back to incubate.
>
> New nest – great.

on the head were important, particularly the pattern of the dark markings on the top of the head and the shape of the band through the eye, as was the colour above and below the eye. The shape and extent of the brown patterning on the throat and the breast was often very useful for identification. Some of the birds were colour ringed, so it was possible to check that the plumage characteristics were indeed diagnostic, and the data allowed us, from an early stage, to build up individual histories for these birds.

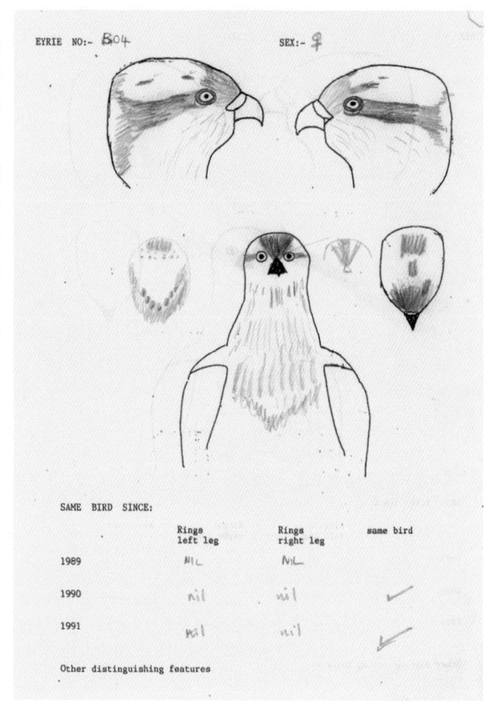
EYRIE NO:- B04 SEX:- ♀

SAME BIRD SINCE:

	Rings left leg	Rings right leg	same bird
1989	NIL	NIL	
1990	nil	nil	✓
1991	nil	nil	✓

Other distinguishing features

ID sketch by the author

Colour ring sightings

Once we had started colour-ringing, we began to check as many breeding birds as possible to see if they had colour rings. It is a hard task to get close enough with a good telescope without disturbing the bird. It can also take a long time, as the female may be incubating, keeping her legs well hidden. The only way, sometimes, is to wait for the male to return and eat his fish, offering the chance to check for rings. Then, once he has given his mate her portion of the fish, she will head off to her favourite perch, allowing her legs to be checked in their turn. Ospreys often perch on one leg, keeping the other tucked up among their feathers; it is funny how it always seems to be the one with the colour ring that stays well out of sight. Nowadays, telescopic digital cameras are so good that colour rings can even be read on enlarged photographs of flying ospreys. Email messages with a digital photograph convey immediate news of migrating and fishing ospreys.

Colour-ringed young

I still find it a tremendous thrill to see well known colour-ringed favourites return from their winter in Africa. Some are now more than 15 years of age, a few as old as 25, and the oldest I have known was 28, that knowledge a direct result of our keeping track of their individual identities.

Age at first breeding

Most young ospreys remain in Africa or southern Europe during their first summer but, as I have mentioned, we have now recorded two one-year-old ospreys returning to England. It is probable that the majority of two-year-old ospreys return to the United Kingdom and Europe during the summer and some may go north to stray into the nesting grounds. We have only recorded one case of a two-year-old bird breeding, and that was the young female which returned and bred successfully at Rutland Water in 2003.

In Scotland sub-adult ospreys generally return and are ready to breed for the first time in their third year but it often takes them longer than this to find a nest site and a mate. In 1981 I recorded that the age of first breeding for males was: 3 years (1), 4 years (1) and 5 years (2); while for females, the figures were: 4 years (3), 5 years (1), 6 years (2).

A recent analysis of the known age of birds breeding for the first time in northern Scotland was as follows:

	3 years	4 years	5 years	6 years	7 years
Male	3	12	6	2	2
Female	11	7	8	4	1
Total	14	19	14	5	3

This data is principally from the oldest and most studied populations in Badenoch and Strathspey, Moray and other parts of the Highlands. It clearly shows that many ospreys are not breeding when first mature and capable of doing so. The average age of first breeding for females was 4.25 years and for males, 4.52 years. A detailed study of breeding ospreys in New England, within an expanding population, gave a mean age of first breeding of 3.7 years in 1984, although it was 5.7 years in an established population along the east shore of Chesapeake Bay during that same period.

Nowadays, although the old established colonies are full, there is still much competition among young potential breeders to join. Instead of breeding as soon as they are able to, young birds are delaying and waiting to find a place within a favoured 'colony'. The opportunity usually arises due to the death or non-return of an older bird at an established nest. Each delayed season for an individual means a 10% increased chance of dying before the next breeding attempt. The average annual survival of adults in the Scottish population is approximately 90%. So, a bird failing to breed until its fourth, fifth, sixth or seventh year has a greater chance of dying before breeding successfully. This, in broad terms, means an *additional* chance of dying of 9%, 18%, 27% or 36% for each year respectively.

This means that on average, in a 'colony' of ten pairs of ospreys, there is a requirement for only two new replacements per annum, yet we regularly record many intruders at nests within these areas. Non-breeding intruders often visit nests containing pairs throughout the nesting season. Sometimes these visits can be surprisingly aggressive and can result in broken eggs or even the eviction of the resident female. There have been reports in the past of birds actually being killed during these fights.

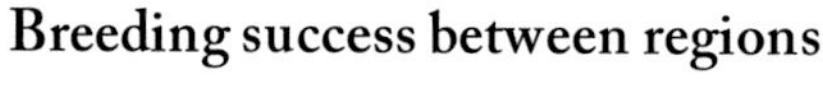

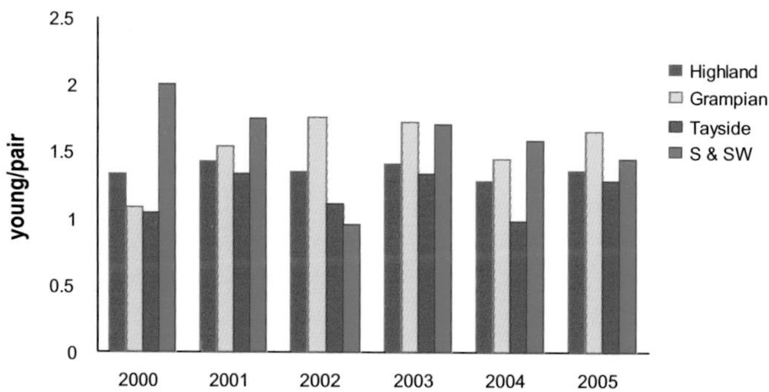

Choice of breeding location

In general, male ospreys return to breed closer to where they were reared than female birds. Our early evidence of this 'natal philopatry' in Scotland came in 1981, when we discovered that of four males one had returned to the very nest where they were reared. The others had bred 3 kilometres, 40 kilometres and 100 kilometres away, while seven females were found at the natal nest, or 5 km, 40 km (2), 50 km (2) and 100 km away from the site. More recent analysis of the sightings of colour-ringed breeding ospreys in Scotland demonstrates the relative natal philopatry of the sexes, as shown in the following table.

Distance and direction of breeding site from natal site

Distance	0–10 km	11–25 km	26–50 km	50+ km
Male	8	9	8	4
Female	2	0	8	24

Direction	Nil	North	East	South	West
Male	10	5	2	7	2
Female	2	10	2	13	3

Adult survival

Although adult ospreys usually come back to the same nest every year, some do move away, so a bird's absence does not necessarily mean that it is dead: it may simply have lost its breeding place. Young, first-time breeders can readily be ousted as can older birds that lose a long-term mate. There is some indication that old 'widows' or 'widowers' can be pushed out by younger pairs of birds and they then wander away. Just such a widowed bird may have been the female who joined the Rutland Water pair in 2001 and 2002. She had a damaged eye and on both occasions two of her three eggs failed to hatch.

To work out adult survival rates, I have looked at four groups of birds. I analysed males and females separately, in two time periods: 1960–1990 and 1991–2004. In each period, I also separated those birds which spent only three years at a nest and compared them with more settled breeders. The table below shows the results:

	Males	Sample size	Females	Sample size
1960–1990	82.4%	18 birds – 102 years	74.6%	32 birds – 226 years
1960–1990 (over 3 years at nest)	87.6%	11 birds – 89 years	87.4%	20 birds – 158 years
1991–2004	83.9%	38 birds – 237 years	82.8%	42 birds – 245 years
1991–2004 (over 3 years at nest)	89.1%	23 birds – 212 years	86.5%	29 birds – 215 years
Survival is probably close to	90%		90%	

Changes of mate and nest

The great majority of breeding ospreys return to breed in the same nest year after year once they have established their 'ownership'. There are a small number of cases in which birds may be ousted from nests in their first year or two of occupation and old individuals may be driven out after the loss of a regular mate that fails to return. Very few birds move willingly from nest to nest but this does sometimes occur. It could be that individuals are moving to 'better' nests within a 'colony', that their mate is late or does not return, or that they are just wanderers. Two females we have studied have bred at three separate nests with different mates:

'White–black EV' bred successfully in Aberdeenshire in 2000, then moved 80 kilometres to nest in Morayshire and reared 3 young. But in the following year, she moved 1.5 kilometres to another nest where she failed to rear any young. She has not returned again.

A Scandinavian female that was hatched in Norway bred in Badenoch for six years and then moved six kilometres away to a new nest which she used for two years. She then moved a further two kilometres in 2004 to another new nest, where she still breeds.

Two males moved locations but only short distances of one kilometre and five kilometres were involved. Two other females studied also moved nests, at distances of six kilometres and twelve kilometres respectively.

Ospreys are monogamous and there are very few cases of polygyny in ospreys in Scotland. There have been two cases recorded where a male mated with two females that both then laid eggs in the same nest. The first was at Loch of the Lowes in 1976 and the other in Angus in 1986. Neither nest was successful in producing young. A male at a nest in Moray apparently was also the mate of a nearby nest which failed. Only a very occasional case has ever been reported abroad.

Longevity and lifetime reproduction

If ospreys survive to adulthood, older birds especially tend to remain faithful to their mate. The longest period I have known is 22 years for a pair in which the male bird had been breeding at the same locality (two different nests) for 25 years. He must have been at least 28 years old while the female had been breeding for 22 years and must have been at least 25 years old. This is an exceptional pair and they have produced 38 young, despite having their eggs stolen in five of those years and having them broken by intruding females in a further two breeding seasons.

At another nest I know, the male partner was 17 and the female 18 years old, and they had been together for 13 years, producing 25 young. The female produced 32 young in total. We have two other records of pairs remaining together over a long period, of 8 years and 13 years.

Variations in breeding success of different pairs and nest sites

There is much variation in the breeding success of individual birds and as members of pairs, and overall breeding success averaged out over a series of years can vary from just over one young per year to over 2.5 young per year for the most successful birds/pairs.

There are variations in the long-term success of individual nest sites, which involves, over time, a succession of different osprey pairs. Loch Garten is a very productive nest, with, over

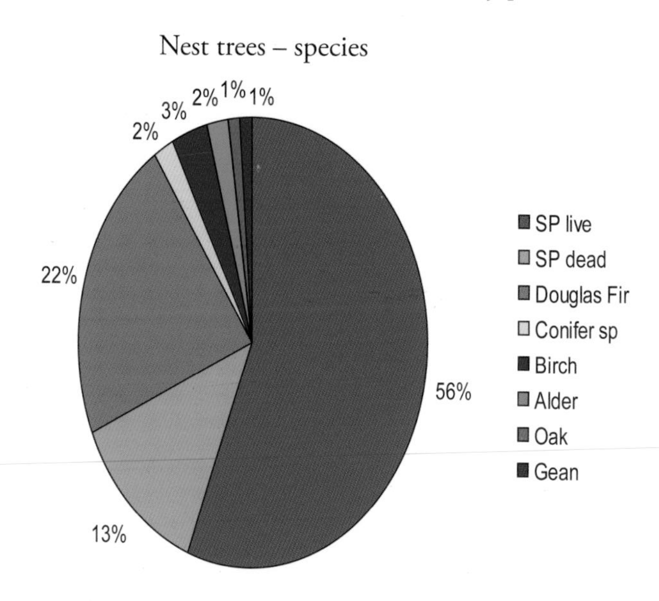

the time span of 49 years, an average of 1.8 young per year, while the poorest performing sites have a productivity of just over one young per year.

There is much variation between different regions, and this will become even more marked as the British Isles are recolonised. Some annual variations are due to differing weather while others are due to fish species and their availability. Coastal and estuarine waters are very rich, but birds using these habitats can also experience high and low peaks probably related to coastal 'haars' (fog) and bad weather, while inland water feeding males seem to be more uniform.

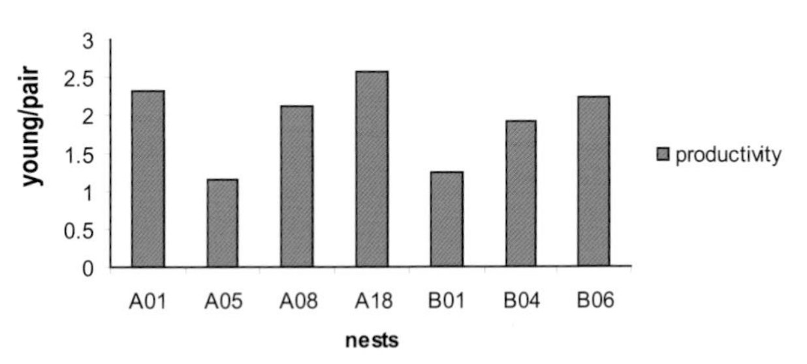

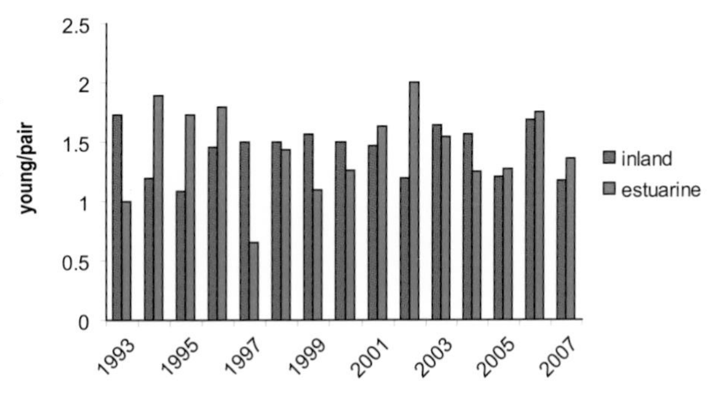

Close-up of young osprey

'Red Z' and 'M99'

I find it very exciting to find 'old faithfuls' whose fortunes I have followed for ten years or more, back at the same nest after another winter in Africa. A favourite osprey of mine was 'red Z'. As my colleagues will testify, I am not keen on naming wildlife, since to me, it is an affront to their wildness and an attempt to humanise them. I always prefer to think of them as wild birds within wild nature. 'Red Z', therefore, was as much of a name as this bird was given. I identified her from the colour ring on her leg as being a bird I had ringed as a chick in 1974 at a remote nest in north Perthshire. She started to breed at nest number 18 in 1980 when she was six years old. She had moved 100 kilometres to this nest from her birthplace and it was occupied by a local male ('M99') that was ten years old. His mate of the past five years had either died or was lost. 'Red Z' laid three eggs and reared two young successfully.

In 1982 she became entangled in baler twine that she had collected from the fields as nest lining. After two days, she was still trailing the twine and, worried that she would get snagged in the branches, I decided to set a trap for her in her nest. Stewart Taylor climbed the tree and installed my trap and went back up to collect her. Once in our hands, I cut away the twine, added the ring 'red Z' and released her back to the nest, where she successfully incubated her eggs. In 1991 she returned once again to breed, in her seventh year, along with her faithful mate who was now 21 years old – our most senior osprey at that time. During their years together, these birds had reared 24 young and were the most successful pair up to that date. They failed that year and the old male did not return in 1992, but his place was taken by a young six-year-old male from a nest about ten kilometres away. 'Red Z', however, returned for a further five years and reared another eight chicks, bringing her personal contribution to 32 young ospreys. She did not return in 1998 and by a strange coincidence, her place was taken by another Perthshire female, aged seven, who, by another coincidence, bore the ring number 'orange Z'.

The male bred for 17 consecutive years and reached 21 years of age; his first mate lived for five years and then he was mated to 'red Z' at the same nest for 12 years. Six of the ringed young from this nest have been recovered as recorded in the table. A female chick fledged in 1996 was found near Dunkeld in 2001, hanging in power cables with a Fenn trap and chain, used for trapping small rodents, clamped to her foot. This poor bird, which died in truly horrific circumstances, may have been breeding in Perthshire.

There have been more colour ring sightings of chicks reared at this nest. A chick fledged in 1983 was briefly seen intruding at Loch Garten on 15 May 1986, while another male chick fledged in 1986 bred near Elgin from 1993 to at least 1998. A chick fledged in 1988 was seen intruding at Loch Garten in April 1991 and a chick from 1994 bred, but failed, near Aviemore in 1997. This bird was also seen at the nearby fish farm in the years 2000, 2001 and 2003.

So, we know that at least nine young survived to two years of age (a 25% success rate) and of course, not all colour-ringed birds will be found or identified at nests. Other chicks fledged from the nest lived to at least 5, 6, 9, 11 and 12 years of age so old 'red Z' and her mate were indeed an important pair of birds for rearing successful young and adding to the osprey population.

Breeding success details for nest in Moray where 'Red Z' bred

Year	Male	Female	Eggs	Young	Comments
1975	M99	new	0	0	New pair built new nest in Moray
1976	same	same	2	1	RSPB's osprey film featured this nest
1977	same	same	2	0	RSPB osprey filmed at this nest
1978	same	same	3	2	Chick shot in Algeria
1979	same	same	3	2	
1980	same	red Z	3	2	New female – 6 years old
1981	same	red Z	3	3	
1982	same	red Z	3	1	'Red Z' tangled, trapped and released
1983	same	red Z	3	3	
1984	same	red Z	3	2	Chick trapped in the Gambia and released
1985	same	red Z	3	3	Chick shot in England
1986	same	red Z	3	2	1980 chick seen at Loch Garten
1987	same	red Z	3	3	BBC's live broadcasts featured this nest
1988	same	red Z	2	2	
1989	same	red Z	3	3	22 young
1990	same	red Z	2	0	1980 chick killed in Guinea
1991	same	red Z	NB†	0	Male 21 years old – 27 young
1992	blue D	red Z	Yes	1	New male 6 years old
1993	blue D	red Z	Yes	0	1986 chick breeding male near Elgin
1994	blue D	red Z	3	2	
1995	blue D	red Z	Yes	3	
1996	blue D	red Z	3	2	
1997	blue D	red Z	3	0	1994 chick breeding near Aviemore
1998	blue D	orange Z	3	2	New female

†NB – non breeding

Red Z

4th May 1982. Nest A10. Miserable wet cold day but started to clear up about 3 p.m. Met Stewart and Richard at the gate and so to the nest at 5.45 p.m. and female incubating. Disturbed her and still badly tangled with orange twine. Put up the ladder and fitted noose trap into nest and changed the eggs. Kept her eggs warm and safe in special box, marked them H82/RHDS (ultraviolet in case of egg thieves). Came away from tree at 6.35 p.m.; female landed in next-door tree and at 6.44 went to sit on eggs and was caught in trap. Stewart climbed back up tree but a real gust of wind and she escaped. Reset trap and at 7.05 p.m. female back to nest; by 7.10 p.m. we had caught her. Female badly tangled with twine round her neck in a slipknot. Cut away the twine around her legs and body. She was already ringed – 'M9965' on left leg; I placed a yellow-red Z colour ring on right leg. Wing 530 mm; bill 34 mm, tarsus 62 mm and tail 245 mm. Released her 7.30 p.m. and she returned to incubate at 7.40 p.m., so put her off again and back up tree and swapped the eggs, and she finally came back to incubate her own eggs after half an hour. Success!

'Red Z' – she was caught to remove baler twine

8

Threats to ospreys

British ospreys may be doing well but they could be doing even better: they still have a long way to go to get back to where they were before the days of persecution. We know that they have difficulty in breeding in new areas and are slow to recolonise places where they once thrived, although this process can be helped by translocations such as at Rutland Water. We also know that about 60 per cent of ospreys die before breeding and that annual mortality among adults is about ten per cent. So what are the threats to our ospreys? How do they die?

Nearly 60% of all reported deaths can be attributed to a known cause, with 23% due either to deliberate capture by humans, accidental capture or natural causes. A further 20% were deaths brought about in some more indirect way by humans, for example, as a result of pollution. Since 1975 there has been a drop in the number of osprey deaths in Europe from shooting but an increase in the number associated with electricity power lines.

In 2007 I checked the cause of death reported for 97 ospreys ringed in the Highlands; 31 were found dead of unknown causes; 8 found dying of unknown causes; 6 from unknown circumstances (e.g. recovery of ring only); 13 shot; 13 hit overhead wires, mainly electricity power lines and suffered broken wings – found dying/dead; 3 found with broken wings; 4 electrocuted by overhead power lines; 7 caught in fishing net – found dying; 5 killed by humans; 3 drowned at sea, and other individual cases killed by train, car, otter and by an eagle.

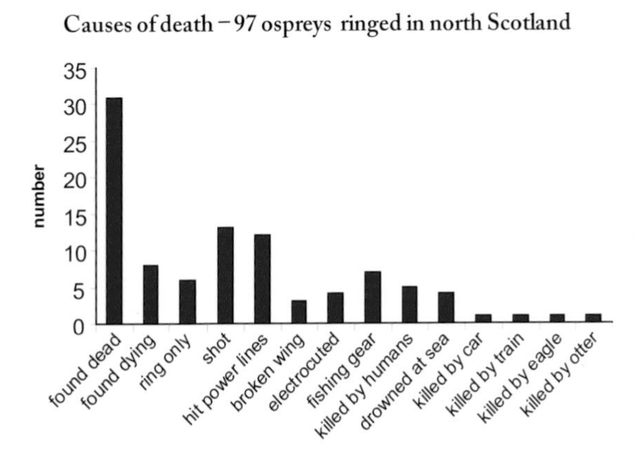

Natural causes of mortality

The weather

The weather has a considerable influence on ospreys both in terms of individual survival and of breeding success. High winds and gales can destroy nests containing eggs or young, particularly those of first-time breeders who have built flimsy nests. Winter storms can also destroy trees and nests, meaning that returning adults have to invest breeding time in selecting a site and building a new nest, sometimes increasing their failure rate. Heavy rain is a particular problem, especially when prolonged over several days, as it makes it much more difficult for the male osprey to catch fish. His plumage gets wet, making him a less efficient hunter, and he finds it harder to see his prey, particularly when water levels rise rapidly and rivers become turbid. Back in the nest, the females find it difficult to keep their chicks warm and dry and, faced with the shortage of fish, the young become hungry and cold and eventually succumb. In some years, such as 1977, mortality due to wet weather can be very high.

Nest F06 before storm damage, Sutherland

Weather also affects the migration of ospreys and our studies have shown that young ospreys on their first journey to Africa can be blown off course and become so exhausted that, unable to make landfall, they die in the sea. It is also likely that if the young are not well fed and in the best possible condition when they set off, the long journey to Africa, especially if the weather is unkind, will be too much for them. The arduous journey for some of them over the Sahara can only be made worse by strong headwinds or dust storms. Even the adults are affected, particularly those returning in spring, hurrying to get back to their nest sites. Losses due to weather during migration are probably higher than recorded, because we clearly do not receive ring recoveries from ospreys dying at sea.

Natural predation

In general, ospreys, as large birds of prey, are highly capable of protecting their nest sites. I well remember seeing a pair of ospreys drive off a young golden eagle which had wandered within half a mile of the Loch Garten nest. They hounded it with violent diving attacks until it turned quickly and headed off.

Re-building F06 after the winter gales, April 1999

Damaged egg

Ospreys are unlikely to be killed as adults, although, occasionally, birds in northern Europe are killed by eagle owls or goshawks, while we have a record of one of our birds being killed by an eagle in Yugoslavia. Young in nests are more at risk, although their patterned cryptic plumage does protect them. The goshawk is a potential predator, and predation may have occurred already in our country, while in mainland Europe eagle owls also kill chicks. After fledging, when still inexperienced, they are at risk to similar threats. Great horned owls sometimes kill ospreys in North America. Red fox often scavenge fish scraps from below occupied nests, and would probably kill any young which fell out.

In Scotland we have had some clutches of eggs raided by pine martens: unsurprising, given that they have learned that the burrowed-out base of a large osprey nest makes an excellent den. There have been some cases in Scotland of common buzzards driving out ospreys breeding close to their nest, while at Loch Garten, a pair of tawny owls, breeding in a nearby nest box, harassed the nesting ospreys at night. This harassment was probably the reason why only one young osprey survived that year. Although crows often live near ospreys in order to scrounge fish scraps, we have never recorded them taking eggs but in Corsica, there has been the occasional loss of eggs and chicks to ravens.

Where ospreys nest on the ground and on cliffs, the number of potential predators increases: red fox, for example, have preyed upon nests in Corsica. In North America, racoons, being excellent tree climbers, are a major predator of eggs. One osprey colony lost a third of its eggs to racoons until tree guards were installed to protect the nests. Ground nesting pairs in the Middle East lose eggs and young to mongoose, while in Australia, such nests are raided by goannas and possums. In their winter quarters, Scottish ospreys come up against a whole new range of predators, with several birds, including a ringed bird from Scotland, killed in Africa by crocodiles. Both the osprey and its ring were found inside the animal after it, in turn, was killed.

Interspecies aggression

As we have established, ospreys often fight over nest sites. Young ospreys ready to breed prefer to take over an established nest and, ideally, to adopt the mate in residence after one of a pair has died. This can mean that several males, perhaps as many as four, will fight over a female, sometimes very aggressively, the birds twisting and turning through the skies. Even more aggressive are the battles fought by females fighting for occupation of a nest. In addition to these fights over nest ownership, intruding ospreys will often closely approach occupied nests. The most damaging intrusions occur when a determined female dives and attacks an incubating mother. This can result in broken eggs and, very rarely, the intruder may even successfully drive off a less dominant, sitting female. In 2004 the RSPB captured on closed-circuit TV the most amazing attack at the Loch Garten nest, in which an osprey laid an egg that was broken during the fight. Talons and bills tore and slashed, and it was astonishing that neither of the birds was killed.

Reports of ospreys fighting to the death seem to me highly credible. Less dramatically, these fights are certainly a cause of nesting failure, particularly in high-density colonies, and, as population density increases, the males have to spend more time at the nests to protect their mates. This leaves them less time to fish, so food shortages may compound the problem and contribute to failures.

Parasites and diseases

There seem to be few recorded cases of ospreys dying from disease or carrying a burden of parasites serious enough to kill them. When ringing, we have sometimes found young ospreys in very poor condition, occasionally wheezing from their nostrils as though they had a bad cold. On occasion, while collecting runt chicks for translocation, they have been found to have serious *salmonella* infections, which later proved fatal. We subsequently had all the young ospreys

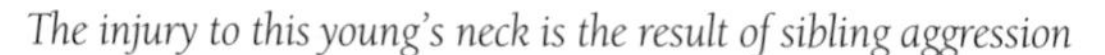

The injury to this young's neck is the result of sibling aggression

involved in the translocation programme to Rutland Water checked by a vet for such problems, and we found no disease other than the occasional infestation by feather lice.

On a couple of occasions, we have seen young in nests with abnormalities. These include stunted flight feathers, which have failed to grow in the sheath, crossed bills and one ringer reported a chick with three legs and feet.

Non-natural causes of mortality

Chemical contamination

Ospreys are at the top of the aquatic food chain so their breeding success and survival can be adversely affected by toxic chemicals in the environment. In North America, catastrophic breeding failure and consequent population decline in ospreys occurred from 1950 onwards due to the widespread use of DDT to control mosquitoes. Residues of DDT accumulated in food chains, including the fish that were the prey of the ospreys. The residues became concentrated in the birds, interfering with their ability to lay normal viable eggs. The eggshells became very thin and were easily broken, dented, or crushed. It was many years before the effects of DDT on wild birds were properly understood, during which time breeding numbers of ospreys declined by 50–90% in parts of their range. Controls on the use of the pesticide resulted in a gradual recovery of the population and in nesting success from the 1970s onwards, aided in some cases by local reintroductions of the birds.

In Scandinavia the effects were less severe but levels of DDT and mercury were high in the 1960s and early 1970s, resulting in a population decline. This appears to have been the case elsewhere in parts of mainland Europe where the effects of DDT may have been more marked because there were additional pressures due to persecution. By the late 1970s, the use of DDT had been phased out and, since then, breeding success has returned to pre-pesticide rates. In Scotland there were few ospreys present during that period to be affected but some of the early pairs did suffer failures and traces of toxic chemicals were identified in eggs that failed to hatch. Other contaminants including dieldrin, mercury and PCBs have been identified in eggs but, in Scotland, none of these have been in high concentrations. While pesticide use is much lower nowadays in Britain, there have been increases in its use in Africa, where the birds spend the winter. This is a possible cause for concern although, thankfully, so far we have no evidence that it has caused a problem.

In Scotland five eggs which failed to hatch in the years 1978 and 1979 were analysed by a Department of Agriculture laboratory in Edinburgh: they had levels of DDE between 2.45 and 13.38 ppm (parts per million); 1.6–14.9 ppm of PCB, and 0.09–0.43 ppm of mercury. A young bird which died in December 1980 contained 7.0 ppm of DDE, 54 ppm of PCB and 0.6ppm of mercury, and a two-year-old Swedish ringed bird found dead in Caithness in 1978 had 5.2 ppm of DDE, 15.1 ppm of PCB and 10.8 ppm of mercury.

Shooting and hunting

In the past ospreys were regularly shot in Europe, sometimes just for fun, and in some places, like Malta and Sicily, this, sadly, is still the case. However, in many countries where the birds were once victimised, the situation has completely changed and the osprey is now left in peace. In some countries birds are still shot because of their depredations on fish farms, even though they are protected.

Scottish birds migrate through Britain, France and the Iberian Peninsular and so on to West Africa. When I examined recoveries in 1991, I found that of 14 birds reported dead before 1980, at least 9 (64%) had been directly shot or killed by man. In the following decade, though, only 5 (22%) out of 23 had been shot. Ospreys are rarely shot in West Africa although there is some evidence that they are hunted for food in Guinea, while young ospreys and eggs are known to have been eaten by people in the Cape Verde islands. Ospreys were also regularly trapped in the past at fish farms to prevent their predation on fish stocks but this practice is now much reduced, even in Eastern Europe where it was once commonplace.

The unluckiest osprey I have known was found in 2004. This was a young bird found shot near Salisbury on 1 September. It was taken to the Hawk Conservancy Trust near Andover where, after nearly a month in care, it was restored to good health. The Trust asked for my advice and whether I would fit a satellite transmitter to the bird before they released it. I visited the centre on 24 September and, finding the osprey in really good health, I fitted the transmitter. It was released on a river not far from Southampton where it spent about a week before flying south to the Isle of Wight. It was so exciting for the people who had nursed it to know that 'their' bird had started its migration again. The next report from the radio was in north-western Spain on 6 October when the osprey was on an estuary in Galicia, a place that I had once visited and knew to be ideal for fishing. How tragic it was to then receive a message a few days later to say that the poor bird had been found dead, shot by a hunter on the very first day of the hunting season.

Rehabilitated osprey in Hampshire – this bird had been shot

Electricity power lines

When I first visited East Germany in the 1980s, I was really interested to see ospreys nesting on the huge metal pylons holding up the power lines. I remember looking down from a hillside and seeing four nests in a line on a series of pylons striding through the forest. On a more recent visit to Cairns in northern Australia, I watched ospreys with young in nests on pylons spaced about half a mile from each other. The pylons provide good nesting sites and in these big structures, the live wires are well separated and so relatively safe. A small number of pairs breed on high pylons in Scotland but the normal design used in the UK is not suitable for ospreys.

Large electricity power lines often cross estuaries, rivers and lakes on migration routes, so ospreys, particularly young birds, are at great risk from collisions as they fly over, or, more seriously, when they dive down to catch fish and fail to see an intervening wire, especially the thinner earth wires. Birds are found dead or with broken wings under power lines, usually having died from starvation. In Germany two males were recorded fighting and inadvertently colliding with cables in the heat of the battle. Both were electrocuted.

In Scotland, in 2007, a two-year-old bird was found dead under pylon lines strung across the River Ness near Inverness, while later in the summer a newly fledged juvenile broke its neck through a diving collision with overhead lines close to a trout fishing pond. The most dangerous situation is when high power lines cross rivers, lakes and estuaries. In these cases, the power companies should mark the earth wires.

Juvenile found dead beneath power cable

Ospreys do sometimes die from electrocution if they land on local distributor poles where the live wires are relatively close to each other and to the metal supports and earths. The level of fatalities depends on the design of the transmission lines in different countries. It appears to be rare in Scotland despite the fact that the design of many local lines is dangerous for raptors, but birds have been killed in this manner in southern Europe. This has been a serious problem in Minorca, with five of seven casualties discovered since 1993 having been electrocuted. There was only a tiny population of ospreys on the island in the first place and, by 2004, the number had dropped from seven to three pairs, highlighting the gravity of the situation. Much work is being done in many countries to minimise the potential risk by changing the configurations of the supports carrying the lines: not only are they dangerous to birds, but accidents and the resulting power cuts pose problems for local communities and the power companies.

Egg theft

It is easy to think of egg collectors as a Victorian phenomenon and it often surprises people that nest robbing is still a problem today. It has been illegal for 50 years and the penalty for stealing an osprey egg is as high as £5,000 per egg, but still it goes on. During the 1970s and 1980s, egg theft was a real problem for ospreys, the thieves going to considerable lengths to get what they wanted. Clandestine visits were made during the night and sometimes the thieves went so far as to cut the protective barbed wire from the trees and burrow through the bottom of the nests to remove the eggs. In 1988 11 out of 49 nests were robbed in Scotland, and 9 out of 49 nests robbed in 1989. All this was done just to add a clutch of eggs to an illegal collection. Contrary to popular belief, there was no money to be made, as the eggs were worthless: the real motivation for collecting was often the perverse thrill of the expedition itself under difficult and illegal circumstances. I often thought it was carried out in order to upset the conservation groups that were trying so hard to help in the recovery of the osprey. I found it particularly disturbing to go back to an osprey nest where I knew the pair had returned and laid, only to find it empty and to discover the all too obvious evidence that someone had climbed up and stolen the eggs. For

A2 robbed by egg thieves – again

Monday, 30th June 1975. Up at 0530. and to the Rothiemurchus nest with Ian – walked out to the tree – male and female present, a few curlews. I climbed the tree at 0630 – signs of climbing irons in the tree and when I got to the top found three hens' eggs in the nest. Robbed. Took one egg down tree. Birds back to the nest. Later back to nest and collected other eggs and found pieces of clothing and twine in barbed wire. Went to the police station in Aviemore and gave a statement etc.

the birds, it was another wasted year for breeding, as ospreys lay a second clutch only on very rare occasions.

I remember being particularly angry on one occasion when the first clutch of four eggs that I had ever discovered were taken just a week or so later, from a nest near Aviemore. In another incident, in 1990, thieves even took the eggs the day before the first one was due to hatch. The police and the RSPB tried to prevent thefts but it was impossible to guard every nest for the whole five weeks of incubation, nor was that a desirable way forward. Sometimes the army helped with protecting the nests, which proved very effective, but the real deterrent has been the introduction of stricter penalties in recent years. If you steal eggs now, you are likely to go to prison. As a result, egg thefts are declining and should soon be a thing of the past.

Over-fishing or loss of fish stocks

Ospreys, of course, rely on fish so over-fishing in any part of their range, or the loss of fish stocks through ecological damage, is liable to harm them. We do not know if this is a problem yet but in recent years the numbers of ospreys fishing in Findhorn Bay in northern Scotland has declined. We do not know why but we believe it must be a reflection of a reduction in the number of flounders. Whether

Two charged after eggs are stolen from osprey nest

Only days before they were expected to be hatched, the eggs of the ospreys at Loch Garten were stolen early yesterday despite a round-the-clock vigil maintained by volunteer wardens attached to the Royal Society for the Protection of Birds.

Inverness-shire police said last night that two youths had been charged with stealing the eggs.

Mr George Waterston, Scottish director of the RSPB, said that at about 1 a.m. yesterday two wardens in an observation post about 150 yards from the eyrie in the now famous Scots pine at Loch Garton heard the alarm cry of one of the parent birds. With the aid of night glasses the wardens saw a figure silhouetted against the sky up the tree. The wardens dashed out to try to catch the intruder but he made off.

EGGS MISSING

Immediately all the wardens at base camp at Boat of Garten were roused and police informed. A search of the area was made, police going from Aviemore, Carrbridge and Boat of Garten. Later two young men from England were interviewed by the police.

A senior police officer of the Inverness-hsire Constabulary said later that charges had been preferred in connection with the incident and a report would be sent to the procurator-fiscal. So far the eggs had not been found.

Mr Waterston said that examination showed that all the eggs — probably three — had been taken. They should have hatched in about a week, and already several thousand visitors had been to the observation post about 200 yards from the eyrie, from which, last year, 39,500 people watched the successful rearing of three yuong birds.

People continued to flock to the observation post yesterday unaware of what had happened. They saw the female bird continuing to sit on the nest with the male helping her to repair damage done to the eyrie.

In addition to a 24-hour watch, barbed wire had been strung round the tree and an electronic alarm device had been attached to it. Mr Waterston said he thought that possibly ropes had been used and that there had been a failure of the alarm system.

What would happen now was not immediately clear. The society would probably keep the observation post open meantime as the parent birds might remain in the area and build a "frustration eyrie" in which they would not be able to lay any eggs this season. But Mr Waterston was optimistic that the birds would return next year to try again to raise a family.

Egg collecting problems

17th June 1976. To Dingwall at 1 p.m. for lunch at the National Hotel with the police at invitation of Chief Constable Andrew McClure. Useful dinner; then to police headquarters for meeting and gave lecture to 15 chief superintendents and senior staff – first half on rare birds and then on the law and illegal falconry – very useful discussions.

THE OSPREY OUTRAGE

THERE is, of course, a widespread sense of outrage over the latest act of plunder in the Highlands. Four eggs of the osprey, one of our rarest breeding birds, have been snatched from a secret site on Speyside, the third year running this despicable kind of pillage has been perpetrated. Bird conservationists are particularly incensed about the latest raid because this is the largest number of eggs an osprey has laid since the last century. But anger over this "night hawk" operation, obviously carefully planned and daringly executed, extends beyond the ranks of bird-watchers. All who have concern for the delights of nature and a love of creatures great and small will regard this as a grievous loss.

What of the culprits? It may be that bigger fines, or even tighter security and vigilance, will not deter those fanatic collectors who will go to any lengths to reach their goal. Although eyries are monitored by wardens, it is almost impossible to watch all the sites round the clock. And how is it possible to get the message through to thieves of this calibre that they are as much robbing themselves of honour as gaining anything?

It seems appalling to contemplate, but the record as it stands indicates that the community should brace itself to expect this kind of desecration at regular intervals, as long as there are people around whose enthusiasm for collecting rarities and whose rapacious instincts outweigh their moral sense and respect for the sanctity of life. If society as a whole is to protect itself against such individuals, there may have to be a re-thinking of a wide range of values and a constant re-statement of ecological fundamentals.

It means that much more has to be done to ensure that our heritage is protected, that the meanest type of personal "enrichment" at the expense of the impoverishment of the environment has to be shown up for the vile act it is and that, morally, socially and psychologically, all of us have a duty to become more aware of what we can do daily to ensure that future generations will not come to feel that they have been robbed of glories which cannot be replaced.

Yet more publicity for the wrong reasons as another incident of egg thievery is reported in The Press and Journal

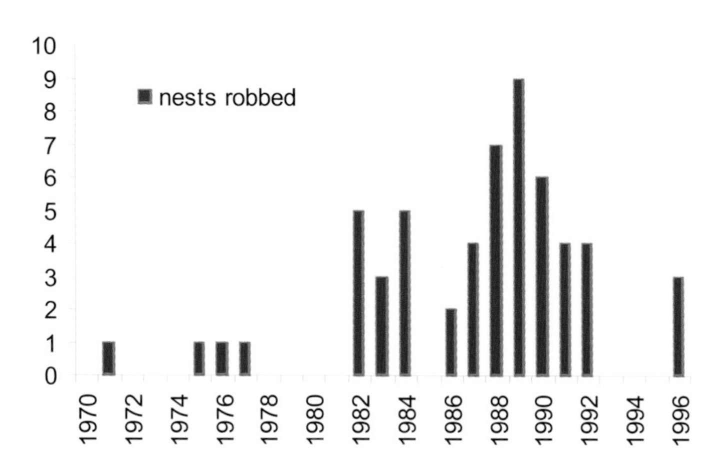

Runt on the left, due to fish shortage

this is due to natural fluctuations in the fish population, changes in the ecology of the bay, possibly including alterations in the nature of discharges, or the impact of the ospreys themselves, we do not know, although, in 2007, numbers of flatfish seem to have recovered.

More seriously, reports of increasing large-scale commercial fishing along the African coast by European trawlers will almost certainly impact upon fish stocks there. This will not only rapidly affect the local people but ultimately the wintering ospreys of Europe as well. It would be a disaster if that were to happen.

The failure by ospreys to recolonise parts of the north and west Highlands may be due in part to a present-day lack of fish in many lochs, due to acid rain. This is a problem that has been exacerbated by long term overgrazing by sheep and deer, and the planting of extensive conifer plantations in the water catchment areas, adversely affecting biological productivity.

Lucky female from A10

5th July 2000. Phone call – Mike at fish farm just found osprey in side netting; 09.50 a.m. I extracted it. Very wet wings and tail. Took her back home and dried out with a hairdryer; it was 'white-black EB' the breeding female from nest A10. Wing 510 mm, weight 1685 g, in fact very fat. Took her back to the nest at 11.45 a.m. and she flew straight to the eyrie – intruder calling at a male osprey overhead. Her two young stood up in the nest and within four minutes of release she was feeding her young from a fish left in the nest by the male. Lucky bird.

Fish farms

Ospreys regularly visit fish farms and, occasionally, individuals may die in the anti-predator nets stretched over the ponds. It is also highly likely that ospreys get caught in monofilament nets placed across fishponds in southern Europe and possibly also in Africa. These nets are nearly invisible and should be replaced with better, visible deterrents: effective netting fitted well above the surface of the water can prevent predation by ospreys and other birds without putting them at any risk.

There may also be cases where ospreys are still deliberately killed for taking farmed fish from ponds. Forty years ago deliberate killing was commonplace. In the autumn of 1953, 98 migrant ospreys were shot at just a few farms in Lower Saxony in Germany, while, in the same era, pole traps killed 20 ospreys at one Mecklenburg fish farm. Thankfully, a more enlightened attitude generally prevails today, but ospreys are probably still killed illegally at fish ponds in Europe.

Fishing line and tackle

Occasionally, but increasingly, when we are visiting osprey nests in the summer to ring the young, we find nylon fishing line, hooks and even bubble floats in the nests. Sometimes the chicks are tangled in nylon line or may even have a hook embedded in their throat, sometimes even in their stomach. Their wings and feet may be entangled and sometimes snagged to branches in the nest. We can cut them free, but otherwise they would certainly starve to death. Adult birds have been found tangled in trees, sometimes just as a remnant skeleton blowing in the wind. Tragically, breeding adults have been lost in this way.

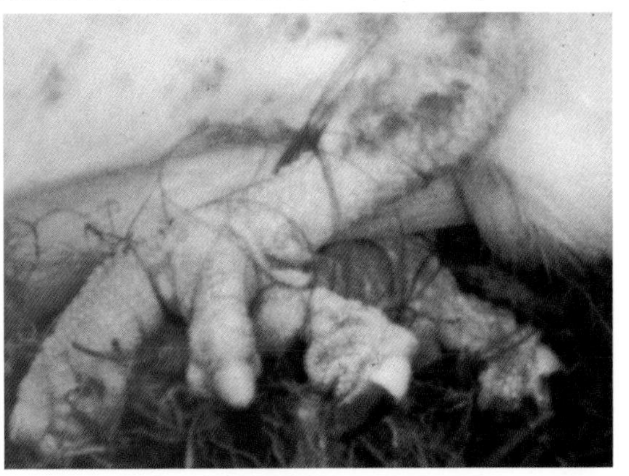

Nylon fishing line around foot, July 2007

Some fishermen do not know better, some are simply irresponsible in discarding line or cutting it free when it becomes snagged, and many are using too light a nylon line in the first place. It is probably a greater problem at venues where novice fishermen are likely to be trying their luck. Sometimes the line breaks and a fish swims off with a hook embedded in its mouth, towing several metres of line and possibly further hooks and floats. If, in the future, an osprey catches that fish and takes it to its nest, the young or even the adults are likely to swallow hooks or become entangled.

I recall an incident involving a single male chick in a nest near Carrbridge. Its feet were completely entangled in nylon line and actually fixed to the base of the nest. I managed to cut it free and take it down to the ground to be ringed, later fitting a satellite transmitter to him. He successfully migrated to Africa. In 1999 one of the young ospreys at Loch Garten had a hook far down in its gullet, with several metres of nylon protruding from its throat. I was unable to reach far enough down the gullet to remove the hook so I cut off the nylon line as far down the throat as possible. The bird continued to grow and flew off on its migration. It was a matter of some satisfaction that three years later, this individual was identified back in Scotland, having survived to re-visit the Loch Garten nest.

Baler twine and courlene netting

Ospreys nesting near farmland often fly out over ploughed fields to pick up material to line their nests. This is often lumps of straw and clumps of dead grass, but occasionally they will pick up other material. Since the late 1970s, farmers have secured hay and straw bales with man-made twine, such as courlene, rather than the original biodegradable string. Nowadays, bales are also wrapped in man-made netting or plastic. Sometimes, when farmers are spreading the dung from cattle courts on the open fields in spring, occasional tangles of courlene twine and netting are dropped as well.

Ospreys sometimes collect these tangles as nest material, and it is not unusual, nowadays, to see baler twine and netting hanging in their eyries. On windy days this material blows around the nest and can be a hazard for ospreys. I have already mentioned 'red Z' whom we saved from strangulation from baler twine in 1982, and we have rescued a few young which have been caught up as well. It is likely that occasional ospreys die from entanglement in this way, and the problem has also been mentioned in other countries including North America.

Human disturbance

Ospreys are easily disturbed by people during breeding. Although it is very rare for them to desert their nests altogether, if the birds are frightened off the nest, they can fail to hatch young, usually because their eggs have become chilled. Nowadays, with some ospreys nesting in full view of people, the birds are becoming much more accustomed to human presence. I know of some ospreys that will not even bother to leave their nest when people get within a hundred metres, but that is highly unusual.

Screaming at intruders

It is, of course, illegal to disturb ospreys intentionally and you are not allowed to go near nesting birds. If the birds call at you with loud 'kew-kew-kew' cries, then you are too close and should walk away. Do not try to sneak up on them: they have superb eyesight and become even more alarmed by people skulking in bushes or woodland. If you have disturbed them, walk away in full view so that the birds will be reassured and return to incubate their eggs.

We have, over the years, had instances of failure due to human disturbance but this has usually been accidental. Forestry workers engaged in tree planting have inadvertently strayed too close to an occupied nest, or a motorbike rally has disturbed the birds. In other countries, modern forestry has been a problem by removing old suitable nesting trees. In Finland 95% of nests are in artificial nests in trees, mainly due to harvesting which does not allow for ancient trees suitable for ospreys. Occasionally, walking trails have been created that take people too close to nests, but the main problem is probably over-enthusiastic birdwatchers or bird photographers trying to get that perfect view or shot.

Wind farms

Increasing numbers of proposals to build wind farms in areas where ospreys breed, hunt or migrate have the potential to disrupt the birds and even to kill them if the blades of the turbines catch the birds in mid flight. A range of other raptor species have been reported as casualties but as yet I have not heard of fatalities among ospreys. This could be due to the relative rarity of ospreys or to the fact that turbines have not yet been erected in important areas for the birds. The best osprey habitat should be protected from wind farm developments.

Injured ospreys

Sometimes injured ospreys, perhaps adults caught in fish farm nets, have come into my care and often all they need is time to dry out and rest before being released. Occasionally, exhausted, lost ospreys have landed on North Sea oil platforms and have been sent on to me from the SSPCA. A few years ago, two birds were sent down from Lerwick in the Shetland Islands, having been picked up exhausted on oil rigs and flown off by helicopter. These were un-ringed young birds and after a week of recuperation at a local centre, they were sent down by plane to Inverness, from where I collected them. I looked after them for a few days and then placed colour rings on their legs before releasing them near Loch Garten. They flew off strongly, enjoying their freedom. However, ospreys fare poorly in captivity, seeming very ill-suited to any form of caged life, and usually die. There are records of tame ospreys but, unlike falcons, they have never proved easy to keep or train.

Author releasing rehabilitated osprey at Loch Garten

Releasing tired ospreys from North Sea oil rigs

17th September 1998. Collected fresh trout from Aviemore then to Inverness – plane from Shetland late – collected two ospreys from SSPCA in Shetland which had landed exhausted on oil rigs in North Sea. Put them in the aviary in the evening with four fresh trout. By morning they had eaten three of the trout. Checked both birds and colour ringed them – one bird flew off very well then headed high towards Loch Garten but the other had a limp in its left wing, so I kept it and it flew off strongly on Sunday 20th when it soared upwards in fresh warm breeze towards Loch Garten and then away past Tore hill.

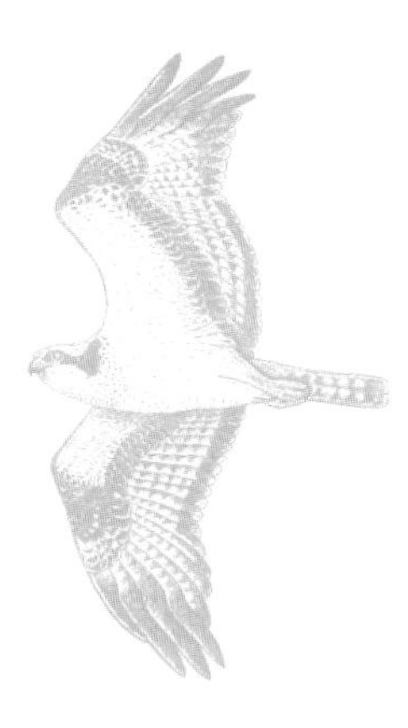

9

Conservation of ospreys

I have always believed in active wildlife management: our environment is so very dominated by man that positive and determined action on our part can help preserve rare birds like ospreys. It is not just a matter of creating nature reserves, designating protection areas or preventing egg thefts, but of giving the birds the best chance possible to produce young. At the moment, osprey numbers are low and we should restore them to all their former haunts and to a population level large enough to withstand the many challenges that they may face. The figures tell us that the number of ospreys in Britain is about a tenth of what it should be. We will know that we have finally been successful when these magnificent birds are once more breeding throughout these islands: along the coasts of Cornwall and Devon, in the fens and marshes of East Anglia and even beside the River Thames in London. It is a real sign of success when a rare species is reclassified from 'endangered' to 'protected'. I am afraid that, for me, rarity is too often an indicator of our own woeful inability to manage a species successfully.

Conservation and protection of ospreys in Scotland

The successful conservation of ospreys in Scotland has been based on trying to ensure that each pair that attempts to nest is left in peace and successfully rears young. The annual monitoring was aimed at checking that ospreys were breeding successfully and identifying any problems and threats.

One of the best ways to protect them in the early years, when they were much more at risk from egg thieves, was to keep the location of the nests as secret as possible and to work with the local people at each nest to ensure their success. At the same time, this strong relationship with private landowners, factors, gamekeepers, stalkers, farmers, crofters, foresters and local folk ensured that their activities, for example forestry and agriculture, close to the nests could be modified during the birds' breeding season. Any of my requests were always met with the keenest desire to help protect the ospreys.

From the earliest days, I was able to reassure the local people that we would keep the sites confidential and that they would not be overrun with birdwatchers. This trust was an important part of the successful protection of ospreys, and was an important part of my job when I worked with the RSPB and also in the last two decades while self-employed. The fact that we used a code for each nest, which individuals could identify for themselves in the annual osprey newsletters, helped forge this partnership.

B01 nest site, 1973

B01 nest site, 2007

In recent years the numbers of pairs nesting closer to public places has increased and there are many opportunities to enjoy watching ospreys at their breeding and fishing sites, including a range of visitor centres and viewing sites (listed in Chapter 12). Ospreys themselves have also become more used to people and less easy to disturb. My view is that their behaviour is changing and that this is a cultural shift. Young which are reared in nests where they can see people are more likely to be able to nest within sight of humans. Our ospreys are becoming more like those on the other side of the Atlantic Ocean, which are much more amenable to humans.

Nevertheless, what should you do if you come across an osprey nest? Firstly, if the ospreys are calling above your head, you are too close and you should retreat and allow them to land back on the nest to continue incubation or to shelter their young. It is sometimes possible to observe them from a distance with binoculars or a telescope, or from within a vehicle without disturbance. If you return and disturb them again, for example, to get a close photograph or show friends, you are breaking the law.

The group that monitors ospreys, including myself, is extremely interested in reports of new osprey nests in order to build up a record of the recolonisation of the UK. Please send details of location (including a map reference), what you have found (occupied nest or old nest), date of finding, and any other useful information on land ownership, and your contact details. Locations are not publicised but contact is maintained with land managers.

Management of nests

Many times over the years, when ospreys were incubating their eggs or looking after young, a sudden storm would blow in and high winds would shake the nests. Looking with binoculars,

Re-building, storm destroyed nests

25th January 1982. Mild grey day but getting colder. To meet Bob McBain the keeper at Skibo. In Bob's vehicle and through the forest to osprey nest which had blown down. Two buzzards and two crows. Nest completely gone and the remains on the ground so climbed up the tree and built completely new osprey nest in very top of the old Scots pine. About two hours to build a really good nest. Then down to Loch Ospisdale and put up two goldeneye boxes on the north side of the loch. The Loch was covered in ice. Put goldeneye boxes on two other lochs and looked for a suitable osprey nest tree on the lower ground but no luck. Lecture to the Sutherland Bird Club in the evening.

A precarious nest, B16, June 2001

you could see the female battened down trying to stay in her nest and keep her eggs warm. Sometimes, we even saw the nest slowly being dismantled by the wind and then it was obvious how important it was for ospreys to build their nests in secure trees; those just placed on the top of spindly trees with little support had less chance of surviving.

We quickly realised that we could stabilise these nests. We became adept at tying in the main base of the nest to solid branches using wire or string. And, soon, we realised that we could anchor these nests much better by cutting the top of the tree down to the first good fork and building a nest so that the following spring the ospreys could return and breed successfully. This also led to us building nests on spec in new areas to encourage new pairs of birds.

Nests that had been in use for many years could also become overshadowed by the trees growing up around them. Ospreys like to nest in the most prominent tree and in order to maintain these old nests, we sometimes cut down surrounding trees to open up the view. Usually, the local forester would carry out this work under our direction in the winter. Another problem is that ospreys often use broken-topped trees, such as Douglas firs, and, over time, side branches below the nest would become the leaders and grow above the nest. If left, the nest would become unsuitable for ospreys and so we would climb to these nests in winter and cut the leader to the level of the nest.

Protection of nests

When I worked at Loch Garten in the early 1960s, one of our great worries was that the eggs would be stolen by egg collectors. The purpose of the RSPB's Operation Osprey was to prevent this happening, so teams of wardens and volunteers were on 24-hour duty to guard the nest. An attack during the day was very unlikely indeed, especially when large numbers of visitors came to enjoy seeing the birds, but night-time was a different matter.

The night watch went on duty at ten o'clock and stayed there until relieved at seven o'clock in the morning. We used to go on in teams of two and take over the log book from the evening watch. The forward hide had the basic necessities: a gas ring, a kettle and the supply of biscuits, tea and coffee. There was also a bunk with blankets and, amazingly, a huge bearskin coat from Russia to keep us warm! The departing team would tell us where the birds were; usually females were incubating with the male in the trees roosting. Looking through the big high-powered binoculars we could just make out the shape of her head against the evening sky.

It was then our job to make certain the eggs were still in the nest next morning. Generally, we would both sit watching through the viewing slits in the front of the hide for the first hour. Often this was a lovely time; curlews and black grouse calling, maybe a fox crossing the moss, or, sometimes, a long-eared owl quartering the edge of the wood. Slowly, darkness would come down and the hoot of a tawny owl would make us jump. One would watch for an hour or two and the other would sleep; then we would change over.

As well as binoculars, we also had other devices. A microphone at the bottom of the tree was linked to a loudspeaker in the hide, which we could turn up to listen for suspicious sounds – and there were many of them. We had a system of wires on the main branches so that if a person climbed the tree, contacts would break and alarm bells would ring. And, of course, there was all the barbed wire draped around the bottom the tree.

Nest A08, branches of the tree have been removed after robbery, July 2000

A08 as it was in the 1970s

It was an effective deterrent but relied on the vigilance of the volunteers and wardens. No eggs were lost during the four years I was warden, but we had many scares. Occasionally, lights would briefly appear behind the tree, turning out to be a car travelling on the road in the far distance. At other times we heard suspicious noises, only to work out it was a fox snuffling round the bottom the tree looking for dead fish or some other innocent sound. Another time I remember hearing very unusual noises, not towards the nest but behind me near the observation post. When I went to investigate, the noise was further away and in the end it proved to be a young couple who had taken one of the fishing boats and was rowing about on Loch Garten in the middle of the night. What I could hear were the oars rubbing in the rowlocks.

Over the years, deterrents were improved with electronic warning devices, lights, pressure pads and infra-red beams, but sometimes the best laid plans of man would be short-circuited by a vole chewing through the cables, red squirrels scampering up trees or red deer grazing below.

What we did know was that one of the most effective deterrents was to make the trees much more difficult to climb, so we cut off all the lower branches close to the trunk. We often wound barbed wire around the trunk and the first branches. Later we used razor wire but today, we have to put up a notice warning egg thieves that they might hurt themselves. A few times we managed to get Army volunteers to help protect some of the nests. But whenever I looked at these defences and the trees devoid of lower branches, I thought how sad our society was that we had to go to such lengths to protect rare birds.

Legal protection

In the UK, ospreys are listed on Schedule 1 of the Wildlife and Countryside Act 1981. Additional legislation affords them specially protected status and this imposes severe penalties for offences, such as £5000 for theft of an egg or bird, confiscation of equipment including cars and a prison sentence of up to six months. It is also an offence to recklessly disturb them at their nest under the Nature Conservation (Scotland) Act 2004.

In the early years ospreys were protected under the 1954 Bird Protection Act but the fines for stealing eggs were very low and the chances of catching people and getting a conviction were even lower. As the extent of illegal egg collecting was revealed by the RSPB and their Species Protection Department, the matter was dealt with more seriously by government and the courts. The 1987 Wildlife and Countryside Act increased protection and penalties, and made it illegal to have possession of eggs collected after that date.

In the European Union, the osprey is protected under the Wild Birds' Directive and it is under Annex I which lists specially protected species. The birds are also conserved within the Natura 2000 list of protected sites and there are Special Protection Areas (SPAs) for ospreys designated in Scotland. The species is also listed in Appendix II of the Berne Convention.

The UK's SPA suite for osprey contains the nesting sites or the feeding areas used by, on average, 60 pairs. This amounts to about 30% of the British breeding population. The distribution of SPAs closely matches the core range of the species in Scotland. Many of the current nest sites were amongst the first to be occupied following recolonisation in the 1950s. Indeed, several of these sites were traditionally used prior to the human-induced extinctions of the late nineteenth century. The Dornoch Firth and Loch Fleet are included as the most important feeding area for ospreys in Scotland, lying at the northern edge of the species' British range. The other Moray Basin estuaries are also included as multi-species sites important also for feeding ospreys.

Loch Garten nest robbed

17th May 1971. Disaster overnight – woken up at 2.30 a.m. by Joan and Brian from the osprey camp with dreadful news that the nest tree had been climbed. I followed them in my car to the peat track – in record time! Most of the chaps at forward hide but police and four wardens hiding by a car down the Mallachie track. A new car with two egg boxes on the floor behind back seat – one contained two meadow pipit eggs. The female was back sitting on the eyrie at Loch Garten; just before 4 a.m. they came back with the news that two chaps from Hartlepool had been caught. Police took them back to their camp at Docharn. Back to osprey camp then to Boat of Garten to see police. Met Constable Cameron and detectives. Doug came over and we checked nest with the mirror pole – the eggs had gone! So back for ladder and climbed tree and found that all eggs gone. Morning with the police etc. – searched wood at Mallachie Road end. Finally found a torch by the Plantation fence and then footprints where they had jumped the deer fence. Tracked them back to the pylons and across to Croftronan but could not find any container with the eggs (they must have hidden them) – finished at 6 p.m. Search called off – the chaps had been arrested in Hartlepool and 153 eggs found in a box in the garden. What a terrible day!

Robbing of nests

Chapter 4 detailed the history of the osprey in Scotland and the problems of egg collecting and shooting birds for taxidermy in earlier centuries. Nowadays, thank goodness, the latter is no longer with us, but the taking of eggs to adorn illegal private collections has been a real problem.

My first experience was not long after I returned from Fair Isle and had started work as the RSPB's Highland Officer. I was then living near Aviemore and Operation Osprey was part of my duties. That year the nest was robbed at night and my diary for 17 May details that disaster. It was an even greater ordeal for my wife and me, because when we were woken in the middle of the night by banging on the front door, we thought immediately of our youngest son, less than a year old, who was ill in hospital in Inverness.

My next involvement with a nest robbery was in 1975, when I noticed that a pair near Aviemore was still incubating in June; when we checked the nest on the 30th we found three hens' eggs had been substituted for the real ones. The culprits were not traced. Two years later, the same nest was robbed again, this time of a clutch of four eggs. From then on, nest robberies started to occur more often and we, in the RSPB, made great efforts to prevent them in cooperation with the police and volunteers, and a small team of summer roving wardens.

Car registration numbers of suspected egg thieves were circulated, tip-offs of unwelcome visitors were telephoned in, and contacts throughout the Highlands made certain that any suspicious activities near nests were reported to our RSPB office in Munlochy. Nesting time was exceptionally busy, and, as most of the collectors had full-time jobs, our busiest times were over the weekends. We staked out nests, we made spot checks of nests day and night, and we even started to code mark the eggs with an ultraviolet pen.

These activities undoubtedly frightened off many people, and helped save clutches of rare birds' eggs including ospreys'. But there was such a determined group who wanted to collect eggs that robberies continued to increase and, in 1989, the worst year, a quarter of the nests were robbed. What made it worse was that many times the nests that were robbed belonged to experienced pairs which had been nesting for many years; the very ones most likely to produce the best broods of young. We had some successes in the field but often our main task was to record the actual nest robberies in detail. On one occasion, one of our roving wardens found three eggs hidden in a box near the tree within hours of the nest being robbed. The thieves had learnt to hide the eggs nearby and return in the autumn, when they were safe from scrutiny. The eggs were put back in the nest and subsequently one hatched and the lucky chick flew.

Trying to combat egg thieves

Nest A02. Up at 6.40 a.m. – a nice warm sunny day – got ladder onto car – John and I went to Glenmore. Went to nest at 07.15 a.m., female incubating and male perched nearby. Four blackcock lekking on the grass parks. Climbed to nest, tree very shaky. Birds very noisy and the male attacked me with steep stooping. Two eggs, very richly blotched and mottled and rather purplish brown, one very beautiful with white to buff background. Marked the eggs with invisible ink pen 'A2 1980 RHD'. Female back onto the nest at 07.25 a.m. when we were only halfway back to the car. 40 red deer near the nest. Later to Tomintoul and met John Hardy and so to the Peregrine nest in grassy limestone valley – a nice river. Male calling as we approached the nest and female came off a low nest with two eggs in the nest scrape. Marked them '1980–P1–RHD', then back down river and had a talk to the under-keeper on the way out.

Our monitoring data was very important because often the real breakthrough came from England, when staff from the Investigations Department at the headquarters of the RSPB accompanied the police on a raid of an egg collector's house. Then we would get a call telling us that an egg had revealed coded marks when examined with an ultraviolet light and a request asking when and where we had marked it. Or, sometimes, coded data in the egg collection might contain something such as a date, number of eggs, type of tree, or even a photograph, which would allow us to link the eggs in the collection with an illegal activity in Scotland. It was this relentless pursuit by the RSPB of criminal activity and the ever increasing fines and penalties which started to end this unacceptable activity. It was hard for me, having grown up in the late 1940s, to believe that the simple childhood activity of taking an egg from a nest could lead to an obsession to rob whole clutches of the rarest birds year after year.

Evidence left by thieves

13th June 1984. Very early to the Cannich eagle's nest with TV cameraman; pair of eagles flew off the crag; ringed single eaglet chick; one dud egg and blue hare in the nest. In afternoon went to Morayshire to check ospreys. First nest female feeding young – could see it from the road. Went to nest B2 – nothing at the nest so something wrong! Walk to the base of the huge Douglas fir to find 6 inch nails driven into the tree and a blue and red rope hanging from a lower branch with two karabiners – nest had been robbed.

A favourite pair lose their eggs

20th May 1995. Loch Insh. 10.30 a.m. Pair by the nest – one below nest on branch, other on a dead tree. Female goldeneye and five young on loch. 10.45 a.m. in boat and landed on the island – walked to the nest – both flew off with no noise. Tree had been climbed with single spikes – old pair of binoculars left under tree – eggs robbed. Danny saw them on the nest Thursday morning and Ross noted birds off on Thursday afternoon and they were off on Friday.

OSPREY TREE —NOW THE ARMY MOVE IN

P&J 1/3/72

CONCENTRATION camp type security measures are being planned for the famous Osprey's eyrie at Loch Garten Bird Sanctuary when the birds return to nest sometime in April.

Last year, the nest was raided and two of the eggs—coveted by ornithologists and thought to be worth up to £100 on the black market—were stolen.

Now, officials of the Royal Society for the Protection of Birds, determined there should not be a re-occurrence of the theft, are to re-think completely their security arrangements at the sanctuary.

They will be helped by Army 'know how.' Men of the 2nd Batt. 51st Highland Volunteers are to visit the sanctuary on March 5 and by the following day, the 30ft. nesting tree will be completely enclosed by an 8ft. barbed wire barricade.

At the recent trial of two men who were convicted of stealing the two eggs last year, it was stated the present precautions were completely inadequate.

Yesterday, a spokesman for the RSPB said: "The decision to build up security at the sanctuary is a direct result of the recent court case."

Apart from the barbed wire, a number of other things are planned, including a secret alarm system.

Capt. Donald Young, Fochabers, OC HQ 'A' Company, will be in charge of the work.

MEN of the 51st Highland Volunteers erect barbed wire round the base of the osprey tree at Boat of Garten.

TOUGH new defences around the osprey eyrie at Loch Garten were completed yesterday by men of Headquarters Company, 2nd Battalion, 51st Highland Volunteers, from Elgin. When the birds return from their winter in Africa they will find huge new barbed wire entanglements firmly fixed to deep-driven metal stakes encircling their nesting trees.

Army wire

The Army move in to help with the conservation of ospreys in the fight against egg thieves

Warning potential egg thieves

Razor wire on nest tree

As the years went on, I became more and more disgusted by the activities of these people, and, generally, with fellow human beings, when I found another pair of ospreys, which I had seen incubating eggs a few days before, just flying about their empty nest. Only on the rarest occasions did they re-lay, so on these occasions their long hazardous journey back from Africa to breed had usually been futile. I also realised that the egg collectors were undoubtedly getting a vicarious thrill by trying to beat us, and that the more we draped nest trees with razor wire, cut off branches, and carried out watches, the greater the challenge and excitement must have become. I was now convinced that instead of publicising every robbery, we should keep quiet, inform the police and the RSPB Investigations Department so that it could be investigated and the thieves caught and prosecuted, but not give the egg thieves the thrill of seeing their misdemeanours reported immediately in the media.

Robberies are now far fewer, and the numbers of pairs breeding are so much greater. In some recent years, more eggs have been eaten by pine martens than stolen by people. But a few years ago we did have nests robbed in the Aviemore area several years in a row, and this led to Marine Commandos from Arbroath carrying out an exercise in 2000. At a meeting with the police and the Marines, I outlined which were the most threatened nests and the most likely weekend for a raid, as well as briefing them on the behaviour of ospreys. The Marines arrived in force on Friday, 5 May and each of the four teams dug an underground bunker from where they could view the nest. Radios connected them to their base for they would be in hiding until Monday morning. Next day, at nest A13, the Marines caught two prospecting egg thieves at the bottom of the tree and escorted them away to police custody. Success for osprey protection and, as the Royal Marine officer said, great training for the men.

Saving eggs and young

Occasionally, we have managed to save individual eggs and birds, but

Rescuing a chick from the storm

22nd July 1974. Very strong winds overnight and the Rothiemurchus nest was blown out. Jimmy Gordon the keeper rescued the chick at 9.30 a.m. from under the nest. In afternoon, it looked fine to me so Tony, John and I collected the ladders and went to the nest. The female osprey still there and noisy but the nest tree was broken topped – with only just a few sticks left in the crown. I cut off top of tree and made a firm platform of spars, then filled it was sticks, heather, etc. to make a passable looking nest. Male arrived and also flew around calling. Ringed the chick 'M9977' and put it back in our eyrie. Withdrew at 7.30 p.m. after an hour's work. Watched until 8.45 pm but although both birds kept flying over and around the eyrie, and the female collected a few sticks, neither would go to the nest and the chick remained hidden from us. Next morning I was at the nest 6.50 a.m. and female on our eyrie with the youngster – nest looks OK – male arrived 7 a.m. with smallish fish from west: female feeding young for over 15 minutes – everything looks OK.

Rescue operation saves ospreys

THE ROYAL Society for the Protection of Birds revealed yesterday that gales had blown one young osprey from its eyrie on top of the 40ft. pine tree at Loch Garten and had destroyed another eyrie containing one young osprey at an undisclosed site on Speyside.

But an emergency rescue operation by Mr Roy Dennis, the society's Highland officer, and Mr Tony Pickup, the warden at Loch Garten, successfully overcame this near-tragedy.

Mr Pickup said yesterday: "At Loch Garten one of the three young ospreys was blown right out of the eyrie onto the ground. We kept a careful watch and after three days as the parents had done nothing to recover their fledgling, we collected the youngster and put it back in the eyrie with the brood.

"That was in July, and since then all three young ospreys have made excellent progress."

At the other site, he and Mr Dennis rebuilt the shattered eyrie on top of the pine tree as best they could and put the young osprey back. This effort was also successful.

It is now time for the ospreys to leave their Highland homes for the warmer climate of Africa. After Bank Holiday Monday on August 26, the RSPB will close down their viewing post at Loch Garten.

"The female has already gone. The male and his three youngsters, who have all gained their wings, are still around, but I expect they will migrate any day now," ... Mr Pickup.

Osprey rescue, July 1974

Saving an osprey chick

1st July 1997. Really wet day – in fact torrential from early morning – cold as well, only 6°C, rain all day with flooding and also on into night. Showers on 2nd and on-off drizzle on the 3rd. Went to Morayshire with Bob Moncrieff and the Forestry Commission guys. First nest had failed. Two dead chicks – effects of the rain – and a dud egg. Next drove through flood-damaged road to the nest in the gully. The nest tree had been washed out by the big flood in the tiny stream and the tree had fallen into the wood – undermined by flood. Saw it was missing from the car. Found one dead chick with broken wing in the wood – then a live chick and also a one and half pound rainbow trout by base of tree. I fed the chick. Male and female present but very high up and they moved off. Decided to cut the top off 20 ft Sitka spruce close by and we built a nest and lined it with wood rush. I put chick in new nest – female perched well down the stream. (They successfully reared it.) Found the other nest had also lost their chicks to heavy rain – home: despondent due to the flood damage.

really most of the natural losses are due to nature, and it is not necessarily appropriate to intervene. In July 1974 I received a telephone call from Jimmy Gordon, the gamekeeper on the Rothiemurchus estate near Aviemore, telling me that very strong winds overnight had blown out the nest near his home, he had rescued a single chick from under the tree, and it was in a warm box at his home. In the afternoon, the winds had abated so I collected two RSPB colleagues from Loch Garten, examined the chick and ringed it. I decided to give it back to its parents, but the nest was gone. With ladders we got to the top of the tree, I cut off the remains of the broken branch and made a firm platform of wooden spars and built a quick nest. The male flew over once and called, but drifted back towards the wood. I placed the chick in the nest and we withdrew at 7.30 p.m. after an hour's work. We watched for 45 minutes, but although both birds kept flying over the area neither would go to the nest and the chick remained hidden.

I was back there at 6.50 a.m. to find the female on the nest with her young and the male arrived two minutes later with a small fish. I watched the female feeding the chick for 15 minutes and then left. In August this youngster fledged, job well done.

In 1995 Keith Brockie and Jason Godfrey rescued two surviving young and placed them in a re-built nest, after a storm destroyed the nest containing three chicks. In 1999 floods in Moray washed out a nest tree and it fell into dense spruces; when I visited the site two days later, I found one dead chick and a survivor. I built a temporary nest in a low spruce, and the parents returned and reared the young. In the winter, we cut this low tree and built a good nest in a high tree, which has been successful ever since.

Ospreys are surprisingly easy birds to work with and will accept re-built nests and even translocated young. Of course, it requires great knowledge and experience of the species, and also, nowadays, the necessary licences. Recently, in Spain, a pair of birds with eggs which failed to hatch have been given small young, translocated from Germany, and they reared them as if they were their own.

In the past we have even had some success with artificially incubating eggs which had been abandoned. In the first case, in 1977, a forest fire raged close to a nest in Strathspey and the birds were prevented from incubating the eggs. Unfortunately, conditions were so dry the fire got into the peat surrounding the nesting area and was impossible to put out for many days. We decided to remove the eggs, and Doug Weir placed them in an incubator at the Highland Wild-life Park. One egg hatched and the chick was put back in the nest as a substitute for dummy eggs which had kept the birds incubating.

In 1985 the famous Loch Garten ospreys also had similar problems when the male sustained a damaged wing on 13 May and later died; this meant no fish for the female and so two days later she deserted her nest. We decided to try to artificially incubate her three eggs; two hatched and one survived and was fostered in another osprey nest, where it was successfully adopted and reared to fledging. Sadly, later in the season, the female's new mate tragically died by getting entangled in netting at a fish farm.

Over the years, various ospreys have come to me either injured or weak. With luck, those that are just weak, usually wet and hungry, can be quickly dried out, warmed up and fed with fish, and usually within a few days are fit to be released. Ospreys with broken wings are very difficult to rehabilitate, despite veterinary attention. Most problems are associated with the fact that they are usually found days after sustaining injuries by which time it is too late to help.

Conservation on migration and in wintering grounds

Our recent migration studies, using satellite telemetry, have shown us in greater detail the importance of stop-over sites, where ospreys feed up during their migration. Sometimes these places are already recognised as important for nature conservation, or may even be designated as protected sites. Now we can add an extra importance for their conservation and management because they are essential feeding areas for ospreys on migration. It is important to protect these special sites and to minimise dangers to ospreys, particularly from power lines stretching over or near water.

All of the British ospreys, in fact all of the European ones as well, winter in Africa, although a very small but increasing number winter in Portugal and Spain. It is clear that there are important wetlands in West Africa, where more than usual of our Scottish ospreys winter. Some of these places are already protected areas, often because of waterfowl and waders, and are a focus of the migratory waterfowl convention between Europe and Africa. Hopefully soon there will be a similar convention for migratory raptors. Northern countries need to help with funding and management for nature conservation in Africa. Another urgent problem is the ever-increasing pressure placed on coastal fisheries by European fishing boats. This is undoubtedly a serious issue for local people, but it is also in the same seas that many of our ospreys fish through the winter.

Natural recolonisation

Ospreys throughout the world are very slow to recover lost range. This has been most obvious in North America where their recovery from the pesticide crash of the 1950s and 1960s was very slow. Conservationists have overcome this slow natural recovery by translocating and rearing young in hacking cages and by building artificial nests. In Scotland too, the bird's recovery has been very slow, with an average natural spread of about four kilometres per year.

This slow progress is due to various factors. Ospreys prefer to breed near other ospreys as they are semi-colonial, fishing at common feeding sites and usually faithful to their eyrie and to their mate. Males have a strong natal philopatry with females being more widely distributed. Young adult ospreys prefer to take over an established eyrie in a 'colony' or build nearby, rather than build a new eyrie in a new locality. Ideally they prefer to find an experienced breeding bird with an established nest which has lost its previous year's mate.

Of course, pioneering pairs do establish in new localities, rarely more than 50 kilometres from the existing breeding range. Experience has shown that it is important that they are joined by other pairs within the next few years to create a new colony; if they are not, the colonisation attempt may fail. The growth of each new colony is slow at the outset but, if successful, it becomes rapid and then levels out. In some cases it may then even decrease.

As mentioned in Chapter 7, many ospreys do not breed when they are capable of doing so, waiting instead until they have found a place in an established colony. Intruders can disrupt the breeding attempts of established pairs and even fight them to the death. In these circumstances, it is good management practice to try to encourage ospreys to breed in new regions of the British

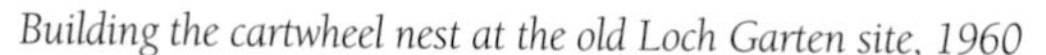

Building the cartwheel nest at the old Loch Garten site, 1960

Isles by providing secure eyries and, where appropriate, by translocating young. (The Rutland Water Translocation Project is discussed further in Chapter 10.) In fact, the increase of large water reservoirs throughout the world in the last century has benefited ospreys, and interestingly in North America the recovery of the beaver has created habitats for ospreys.

Building and re-building nests

I built my first artificial nest in Scotland in 1960 using an old cartwheel. It was never used and, looking back, I am not surprised, since I did not think to add any nesting material! In the last 40 years, though, as the osprey population has increased, we have become much more actively involved in nest management. We started out by repairing nests that had become damaged by storms, having discovered that this was an effective way of increasing the population and enhancing breeding success. Ospreys can waste valuable time early in the season attempting to re-build damaged nests, especially when a main supporting branch has broken off. This can result in nesting failures and even to the break-up of breeding pairs. Carrying out nest repairs is real, practical conservation work and is appreciated not only by the ospreys themselves but also by local people. And it is such a pleasure to see a nest that you have built being used.

In summary, we build nests for the following reasons:

- To enhance breeding success of established pairs by re-building eyries damaged by winter gales or those which are insecure.
- To maximise the production of young while the population is depleted.
- To move pairs from robbed or disturbed nests to a new nest in a secure area.
- To provide secure eyries for young pairs which arrive late and may not build a nest in time to lay eggs. This encourages young birds to breed successfully at three years of age and can increase lifetime reproduction. First-time breeders are the most likely to lose nests in summer storms.
- To encourage ospreys to spread out from the nucleus of a 'colony'.
- To encourage ospreys to join a breeding pair in a new area and thus create a new 'colony'.
- To encourage ospreys to breed in a new region.
- To provide nest sites in areas with few suitable natural trees, especially in large areas of new afforestation where ospreys are clearly restricted by lack of nest sites, or in treeless wetlands.

Of course, ospreys can build their own nests but while they are rare, and still missing from much of their ancestral range, active management can only be beneficial to the species. Nests have been built for all the reasons listed above to help with the restoration of depleted osprey breeding populations and to increase breeding success not only in Scotland, but also in North America and Europe.

Pairs breeding in built or re-built nests have been shown to be more successful than those breeding in natural nests. This is particularly the case with first time breeding pairs in nests they have built themselves. It is important to bear in mind that in ancient times, before human persecution, most young ospreys would have bred in long-established and secure eyries. In the British Isles, with its increasing osprey population, we are currently in the process of restoring the suite of established eyries.

The aim is to increase the breeding population of ospreys in the British Isles and, where practical, to restore it to its previous range. A larger geographical spread of the species will enhance the osprey's long-term chances of survival.

Nest building

In the past 40 years, as the population has increased, conservationists across Scotland have become much more involved in osprey nest management, primarily by repairing nests which have been damaged by storms, and by providing new nests. As the years have gone by, we have attempted to make our man-made nests look as natural as possible for aesthetic reasons.

The following table shows the details for presently and recently used osprey nests in the Scottish Highlands. Twenty-three per cent of used nests were built by us and 14% were re-built by us after the destruction of a natural nest through poor weather or tree failure. Similar information is known for other regions and demonstrates that building and re-building eyries is effective wildlife management.

Details of osprey nests in the Scottish Highlands

Tree Species	Live or dead	Natural nest	Built nest	Re-built nest	Totals
Scots pine	live	26	24	3	53
Scots pine	dead	10	2		12
Douglas fir	live	16	1	4	21
Silver fir	live		1		1
Redwood	live	1			1
Silver birch	live	2	1		3
Alder	live	2			2
Oak	live	1			1
Gean (Cherry)	live		1		1
Electricity pylon		1			1
Nesting pole			1		1
Totals		63 (55%)	35 (45%)	17 (15%)	114

Data from Roy Dennis and Colin Crooke (Highland Foundation for Wildlife and RSPB)

Choice of location

The key requirements are:

- adequate fishing areas and breeding habitat
- sympathetic landowners with suitable trees and land
- secluded locations safe from excessive human disturbance

There are four different types of location suitable for building osprey nests:

- within a present 'colony'
- within 20–30 kilometres of a present population
- within 1–10 kilometres of the pioneer pair in a new area
- in a completely new region with no breeding ospreys

It is important to assess the suitability of an area for breeding ospreys before a programme of nest building is started. Nowadays, anywhere in the British Isles is, in theory, possible but there will be a far higher chance of success if the area has a variety of good fishing habitats and available fish species. Rivers, large streams, lakes, lochs, large ponds, gravel pits, estuaries and shallow coastline will all provide opportunities for ospreys to fish. A mixture of these habitats is most advantageous.

Generally speaking, the higher the number of fish species available, the greater the opportunities for ospreys to hunt, and there are real advantages for those which can exploit estuarine species such as grey mullet and flounder. A wide variety of coarse species in freshwater sites are also important. Salmon are not taken as they are too small when they go to sea and too big when they return to the rivers.

Building nest at Easter Ross with BBC's Blue Peter *presenter, Caron Keating, April 1989*

The best place to start a project is in those areas where ospreys are regularly seen and especially those where they linger or spend part of the summer. Remember that sub-adults, especially two-year-olds, often summer to the south of the breeding range.

Once a suitable area has been located it is necessary to find three to six potential nest sites within a radius of five to ten kilometres of the best fishing sites. Nests are most likely be used in new areas if they are visible to ospreys from regular fishing sites or *en route* between good fishing sites. Remember, ospreys have a remarkable ability to locate nests from the air.

Ospreys are becoming more tolerant of humans. Chicks that fledge from nests in view of people are more able to nest in locations close to humans. Ospreys in remote nests are very sensitive to disturbance and will fly up when anyone comes within as much as 500 metres, or even more, while it is possible to get within a hundred metres of pairs breeding in public places without forcing them off the nest. This is now the case in Scotland and is common in other countries, especially where ospreys have never been persecuted.

Nevertheless, it is better to build nests at least 200 metres, and preferably further, from the nearest habitation or public place and it is best to avoid areas of intensive human use. In the best areas for ospreys, nests and platforms should be placed at least 200 metres apart.

Choice of nest site

There are three types of artificial nest site and each is described below:

- nests built in prominent trees to resemble natural nests
- nests built on poles constructed specially for ospreys
- nests built on pylons, towers and other man-made structures

Safety

It is essential that the highest standards of safety are respected when carrying out osprey nest construction. All equipment and building work should adhere to the necessary risk assessment procedures.

Nature conservation requirements

Anyone building a nest must abide by conservation regulations applying to a particular site and, if ospreys use your nests, you will need the appropriate conservation licences to visit them. In England a special licence is required to repair nests once they are in use by breeding ospreys.

Natural-type tree nest

Select a prominent tree. Osprey eyries, being large and built on the crown of prominent trees, are very visible to other ospreys. Behaviour suggests that young ospreys look out for eyries to occupy. The tree should, therefore, be easily seen by passing ospreys and should be in the sort of place an osprey would choose to perch if it had a fish. If the loch or lake nearby is secluded, then a prominent tree on an island or close to the shore could be chosen, though normally a tree half a mile or more back from the water's edge, on land out of usual public use, is best.

Finding the best tree can take days of searching, while building the nest may take only three hours. Take plenty of time to find the best site, and try to think like an osprey. If you put up a nest in a poor site, the chances are that it will never be used, even in good osprey country.

Scottish ospreys nest in trees ranging in height from 12 to 120 feet – the important feature is that the tree is prominent to its neighbours or is in a prominent clump of trees. It is advantageous to have at least one dead, or dead-topped tree close to the nest to serve as perches for off-duty birds, and from where they can collect nest material in flight. It is probably best to choose a tall conifer (Scots pine or Douglas fir are good) or to use a safe dead tree. Deciduous trees can be used but their general shape means that converting them into a nest site is harder than with tall slender trees like conifers.

Most trees in many areas are not suitable for nesting ospreys. They favour tall trees which have been damaged and have broken tops caused by storms, heavy snow, or lightning strikes. These provide a secure base for their nests.

To build a new nest, at least three people are needed, including a good tree climber. Use an extending ladder, climbing ropes, nylon slings, karabiners, a bushman saw, axe, hammer and nails, flexible wire, courlene twine, plastic sacks and a good knife.

One person climbs the tree and secures himself near the top with climbing slings, just below the topmost branches. The climber also secures a karabiner and a small pulley on a convenient branch and runs a long climbing rope to the ground. The rope is joined so that a continuous loop passes from the base of the tree up through the pulley; allowing the ground party to hoist nest material up to the climber. Sometimes the climber needs to remove some side branches as he climbs the tree to allow the stick bundles to be pulled up later.

The first task is to cut off the top of the tree at the first ring of strong side branches, while colleagues on the ground select nest materials. First of all, find three or four strong branches about five inches thick and 5–6 feet in length – these can be live or dead as long as they are strong. Next, gather bundles of dead sticks up to two inches thick and 2–5 feet in length (preferably barkless conifer sticks which have weathered and have a pale look – this paleness makes the nest obvious to passing ospreys). These should be tied in bundles about a foot thick with nylon slings that can be hitched and unhitched easily from the karabiner on the endless rope. Between four and six bundles will be needed. These must be dead sticks, not cut live.

The next step is to fit the long branches into the fork in a triangle (or square shape) to form the main support. Nail or tie them into the main branches and to the top of the cut off trunk. Secondary supports are then tied or nailed across the main frame. The ground party then pulls up bundles of sticks, starting with the biggest, which the climber builds into a circle at least four feet in diameter. The lower ones are tied by twine or soft wire to the base. They should be worked together and some sticks laid across the middle as well. Keep building up until the nest is at least two feet high – it must look big and bulky.

When you are getting towards the finish, hoist up at least two sackfuls of nest lining, a mix of dead grass, moss and leaf litter or rotted straw. This must be pushed deep into the base of the nest and worked in between the sticks. When the nest is full of dead grass and leaf litter it should be flat. Then take one more bundle of dead sticks, the whitest possible, at least an inch thick but not as long as the ones in the base of the next and arrange these around the edge. Some should be pushed at an angle down through the lining to hold the whole thing together. Then spread one more load of earth or leaf litter to form a nest which looks as though it has been home to a brood of young ospreys. Finally, some people have whitewashed the outer ring of sticks to make the nest more obvious from the air, but pale sticks and dead grass will do. The final nest should look as though it has been made by ospreys.

The last important stage is to create an extremely good perch for ospreys, preferably close to the nest. If possible, leave one of the top branches sticking out from the side of the nest as a perch – in fact it is good to leave a perch beside or just above and to one side of the nest. Remove foliage from this branch. Trim a nearby tall tree, or select a dead tree, to act as an off duty perch. If necessary, cut and clean a prominent branch at the top of the nearest tall tree.

Osprey nesting pole

In areas where there are no suitable trees for building a natural looking nest, it is possible to use nesting poles. These have been used to good effect in various parts of the world, the best example in this country being at Rutland Water nature reserve. Many such nesting poles have been used in North America.

First, you need a strong pole, about 30 feet or more in length. The best are those used by electricity companies for local supply lines and it is sometimes possible to beg or buy a few when they are replacing old ones. These poles are very strong and very well treated. Next, you need to construct a square wooden platform to fit on the pole, big enough to hold a nest. Use six-inch by two-inch treated timber to make a wooden frame of 1 to 1.25 metres square. The four inner spars are jointed to form a hole which fits the top of the pole. The outside frame is made of similar timber and the whole frame is covered with wire mesh. Wooden or metal pins are inserted in the outer frame to prevent the nest from falling. The whole frame is fixed to the top of the pole with bolts. Instead of a wooden frame to hold the nest, it is also possible to use a metal basket made especially to fit the top of the pole.

The nest pole is put up in a suitable area for breeding ospreys, embedded at least six feet into the ground using an excavator. The nest platform should be at least 24 feet above the ground. A successful technique utilised abroad has been to site poles in areas which are about to be flooded or, as happened at Rutland Water, to put them up when reservoirs are at very low water levels. Once the nest pole is secure, a nest similar to that described above should be built on top of the frame and tied into the structure. This can best be done by using a safe ladder tied to the top of the platform. Try to make the nest as large as possible. Finally, fit a tall piece of timber, nailed to the pole and up against the edge of the platform, to provide a T-shaped perch several feet above and to the side of the nest.

Artificial structure

Ospreys regularly nest on the top of electricity pylons in various parts of the world; it is common in Germany, Australia and North America, and there are at least four such nests in Scotland. This can sometimes cause problems for the electricity companies, and they have designed special metal baskets, both for ospreys and white storks, which can be bolted on top of the pylons to keep the nest clear of the working parts. These can be very successful nests as they are secure from human interference. It might be possible to find companies that would be pleased to help with a conservation project for ospreys. At this height and in these circumstances, it would not be possible to build a big nest, but at least some nesting material should be placed in the basket to encourage a bird to take up residence.

Nesting platforms of a similar style to those found on poles or pylons can be fitted on the top of channel markers, navigation towers or structures in coastal and estuarine areas, but make certain you have permission. There are other potential sites: at least one pair has tried to nest on a mobile phone mast in Scotland and another nested on the chimney stack of a deserted house.

Nest on top of TV aerial, USA

Maintenance

Once built, it is important to maintain nests in good condition. They should be inspected in March to repair any winter damage. Remember that they may not be used for many years but it really is worth keeping them in good condition, just in case. Do not give up hope. I was told by a friend in the Forestry Commission that a nest I built in a Scots pine in Sutherland in 1993 was finally used in 2004 – a wait of eleven years.

In some areas, especially in lowland England, osprey nests built close to the water may be used for nesting by Canada geese. They are quite capable of deterring ospreys from nesting, especially as they may have settled in the nest before ospreys have returned in early April. In Canada they suggest using the convex lid of a dustbin to cover the nest during the winter, leaving it in place until the first ospreys are seen in April. The smooth surface will prevent Canada geese from nesting earlier. At Rutland Water they have simply draped a plastic cover over the nest to the same effect, removing it as soon as the ospreys arrive back.

It is important to realise that many artificial nests may never be used but it must be worth trying to encourage ospreys to return to their original range throughout the British Isles. It is incredibly exciting to see your first nest used and there has never been a better time to try. With the new outlying pairs at Rutland Water and in Wales as well as, I hope, further translocation projects, the future for this wonderful bird looks very hopeful. My website (which can be found at: www.roydennis.org) provides the latest information on nest building.

10
The return of the osprey to England and Wales

England

By the 1980s the osprey population had increased in Scotland and the possibility that they might one day nest again in England started to occupy people's minds. On visits south, my friends would ask about building artificial nests and whether I thought birds would actually stop and use them. Their hopes would usually have been raised by the sight of single ospreys spending a week or a month of the summertime on some southern lake. Sadly, I usually disappointed them by saying that these were nearly always young ospreys, probably two-year-old birds, and that once adult, they would bypass England and head north to breed in Scotland.

Rutland Water Osprey Project

One of the most exciting and worthwhile projects with which I have been involved in the last decade has been the translocation of young ospreys to Rutland Water, and the subsequent re-establishment of breeding ospreys in

Building artificial nests for Tim

17th March 1995. Rutland Water. Really wild days with gale force west winds, and rain showers later with hail. Out with Tim and his team to the southern sites – to the first nest – really windy – I got up the pole and tied in the ladder. Then the guys hauling up sticks and grass for me to make a good nest in one hour. Tim arrived with Steven Bolt and a 'Times' photographer. Next to the furthest south pole– very high – tied in ladder and built good nest in an hour including photographs. Back to Tim's cottage for a discussion about the translocation idea. Went down very well. Then round by road to the last site – the guys took the ladder across by boat – got really rough and windy and it was a real struggle to get the ladder up and in the end only managed to put up a few lots of sticks and one bag of grass. Then abandoned it as rain, hail and thunder. Exhausted and wet – back to cottage. Left at 3 p.m. and drove to the Highlands. Very heavy snow showers on way north – black ice and snow drifts in Drumochter – home 11.30 p.m.

Tim Appleton building the Rutland Water cages (photo courtesy of Rutland Osprey Project team)

the English Midlands. Rutland Water, near Oakham, north-east of Leicester, is very well known to birdwatchers: home to the famous British Birdwatching Fair, it is a superb nature reserve, an artificial lake of 1255 hectares created in 1975 for water storage and supply. At the west end is a man-made nature reserve of 182 hectares managed by the Leicestershire and Rutland Trust for Nature Conservation on behalf of the owners, Anglian Water plc. The reserve was created by the manager, Tim Appleton, and his staff, and includes lagoons, marshy bays, grasslands and woodland with a rich variety of birds and other wildlife. Rutland Water is a Grade One Site of Special Scientific Interest as well as a Special Protection Area (SPA) and a Ramsar site. Bird hides and trails provide the most excellent bird watching opportunities.

Migrant ospreys had been seen at Rutland Water since it was created, as it provides a rich supply of fish. I first met Tim Appleton in the early 1980s, when he asked me to build an artificial osprey nest in a big tree overlooking the lake. That nest, though, was never used. In 1994, however, two ospreys spent the summer at Rutland Water and soon Tim was back in touch, wondering how the birds could be encouraged to return. So it was that, in March 1995, I returned to the reserve to help erect five artificial eyries, built on platforms mounted on tall poles. It was while we were building them that I broached the idea of moving young ospreys to Rutland, rather than waiting for the very slim chance of natural colonisation. We thought the site offered ideal conditions for a translocation project and were keen to follow in the footsteps of successful projects that had taken place in America. That evening we discussed the idea with Stephen Bolt, the senior ecologist with Anglian Water, who was very enthusiastic, so we decided we would draw up a formal proposal to translocate young ospreys from Scotland to Rutland Water. The key aim of the project would be to restore the osprey as a breeding species in England and thus extend its breeding range, enhancing the overall chances of survival for this rare bird.

In 1995 neither of the ospreys from the previous summer stayed at Rutland Water: one of them was probably an osprey which briefly viewed the nest platforms on its way north. Now that they were a year older, these two birds naturally headed off to breed in Scotland. Meanwhile we prepared our plans and after exhaustive consultations with the relevant government bodies, Scottish Natural Heritage and English Nature, as well as local interest groups, we finally obtained permission to translocate a limited number of young ospreys from nests in the Scottish Highlands and release them at Rutland Water. Others were not in favour and I even heard comments that if ospreys became common in England, people would not visit osprey visitor centres in Scotland.

View from the back of the Rutland Water cage

We were to start during the summer of 1996. I had already gained expertise in reintroducing young white-tailed eagles and red kites to Scotland, but to gain experience with ospreys, Tim and I visited Minneapolis to learn the American techniques from Mark Martell. This visit encouraged us to believe that our project would be equally successful.

Collecting donor young with Forestry Commission

Back at Rutland Water, things were certainly moving along. Helen Dixon had been appointed as osprey project manager and Tim's team was busy building the cages for releasing the young ospreys. These were modelled on cages used for releasing raptors in Scotland and on those we had seen used for ospreys in North America. Cages six-foot-square were built at the top of twelve-foot-high wooden towers, positioned to face out over the lake. The front and top of each cage was covered in plastic mesh so that the birds could look out and learn the details of the place that was going to become their home. The backs of the cages were made of solid wood, allowing the project staff to work behind them without being seen by the birds. It was crucial not to tame the birds but simply to feed them well, in as natural a way as possible. Each cage had an

Dummies being painted

artificial nest built in the corner and a small door for access. Fresh fish could be dropped into the nest through a hatch and a CCTV camera was installed in each cage, relaying pictures back to a small caravan. The translocation technique involved supplying as much fresh fish as the chicks could eat, just as an efficient father would have done, then allowing the youngsters to absorb their surroundings at Rutland Water, including the position of the rising and setting sun and the configurations of the moon and star systems overhead. The hope was that it would all become familiar as the home to which the young ospreys would eventually return when adults. That, at least, was the theory.

In Scotland I was busy checking for suitable potential donor chicks as I carried out my usual field work with the ospreys. I sought permission from private landowners and the Forestry Commission to collect single chicks from broods of two or three young. All those involved were highly enthusiastic and wanted to see us succeed. The project started in earnest on 6 July 1996 at the first nest near Carrbridge. At 8 a.m., while a camera crew recorded events, I took down three young from the nest and, after ringing them, I put two chicks back but kept the youngest. We next drove to Moray where I collected four more youngsters, each from broods of three, so that, by the time Helen Dixon had arrived in her van from the south, we were already well ahead. The young ospreys were kept in a special aviary at my home and fed on fresh fish. I collected the last two birds on 9 July and set off that evening with Helen to drive the 440 miles south to Rutland overnight, with our precious cargo of seven young ospreys. The cool and dark of night-time was a good time for them to travel, and meant that they would not miss out on daytime feeds. The chicks remained quiet in the travelling boxes throughout the whole of the journey.

We arrived at Tim's house at 6 a.m. to be greeted by a large welcoming party. After an initial examination of the new arrivals, the ospreys were placed in small groups in their new nests inside the cages. The same procedure was repeated during each of the next five summers and, by the end, we moved 64 young ospreys to England. Once the chicks were sitting quietly in their cages they soon started to feed and form themselves into new broods. They established hierarchies just as they would in the wild, with a bossy chick getting the first fish and less dominant ones feeding later. Helen and her team fetched fresh fish and placed it two or three times daily in each nest. The CCTV made it possible to check that all the chicks had fed. With the exception of some runt chicks in the first year, which I had selected in an attempt to save them, all of the translocated young ospreys thrived and were ready for release into the wild by late July.

The next important stage of the operation was the actual release. The

Start of the English reintroduction

6th July 1996. Couldn't sleep overnight as worried about the osprey work – up at 6 a.m. and collected Joe Hayes at 6.45 a.m., met the photographers and film-makers at Boat of Garten, and to the first nest near Carrbridge where started the collection of young ospreys for translocation to Rutland Water. Three young in nest A11 – ringed all three and collected the smallest. Finished filming – all went OK. Took youngster back home and placed in the special aviary – fed it on fresh rainbow trout. Joe and I set off for Moray-shire and met Mike Scott at Fochabers, so to the high tree, nest K06, three big chicks – both adults present and very big nest. The ladder just reached the nest, collected smallest chick. Later, Bob and all of us to the large nest near Forres, three chicks but all rather small so took the biggest. Drove home and put three chicks together in an artificial nest and fed them on trout. Helen arrived from Rutland Water in the van. Marek Bulkovski from Poland and later Iain McLeod from New Hampshire came by and saw birds. A very busy day but a good start to the project.

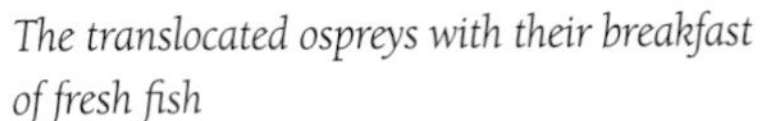

The translocated ospreys with their breakfast of fresh fish

Rutland watch point

young ospreys had by now been flying strongly within the cages. Each year I drove down for the big day of their release and to help with a final check on the youngsters, weighing and measuring them and fitting a tiny radio on a tail feather of each one, allowing Helen and her team to track them. In later years satellite transmitters were likewise fitted to some chicks. Before release, fresh fish supplies were placed in an artificial nest positioned in front of the cages and on a platform that had been built on top.

The watch point with CCTV monitoring of the cages

Translocated ospreys in their cages at Rutland Water

We opened the first cage on 30 July 1996. There were two chicks inside but only one of them – bird number 16 which had been collected in Easter Ross – decided to fly to freedom while the other stayed indoors for an extra day. The front of the cage was lowered at 1353 hours but it was not until 1626 that number 16 made its maiden flight, lasting five minutes, circling over the lagoons before coming to rest on a perch, only to be mobbed by six common terns. This was the most thrilling moment for all of us watching from a safe distance. It was the culmination of a year and a half of planning and work. We felt a mixture of pride and anxiety, a little like parents watching their child disappear behind the school gates for the very first time. Over the following days the rest of the young were set free. Even in subsequent years, release days were always very exciting. Some young would leave quickly while others took a long time to make up their minds to take that final step to freedom. Some made long first flights over the release area while others made quite short journeys, just to the safety of the nearest perch. It never became mere routine and I never lost the thrill of watching the ospreys go.

In 1997 the cages were relocated to a small wooded rise called Lax Hill, which the young ospreys themselves had chosen as their favoured location during 1996. Lax Hill is about 300 metres from the shore and the wood, about 200 metres long by 100 metres wide, and includes a range of mature trees up to 30 metres in height, providing a good variety of perching sites. It also held two artificial nests and was where we had built the first trial nest nearly 20 years before, in our effort to encourage wild ospreys to stay. The cages were erected on the north side, shaded from the midday sun, and the surrounding land was close-cropped sheep pasture. The new release tower held four cages, enough to house twelve young in broods of three.

From 1997 onwards we collected only strong healthy chicks and we selected birds of about six weeks of age, older than the original batch had been. At this age, they were able to feed themselves within the cages. In order to get the birds accustomed to the sight of humans, which ospreys breeding in England would need to be, observers were asked to walk across the front of the pens, at a hundred metres or more distance, two or three times a day.

After the birds were released, visual and radio contact was maintained from an observation post 300 metres in front of the cages, as well as from the main reserve building. These were great times for the observers as they watched the ospreys gaining in strength and flying ability. Soon they were flying out over the reservoir and high above the woods before coming back to Lax Hill. Just as in the wild, the young ospreys learned to collect fresh trout provided in the nests and then flew to favourite feeding perches. Later, they would preen their feathers and sometimes fly down to the water to wash. A few started to dive into the water and a couple were successful at catching fish before they set off on migration.

In late August the first ones left and by mid-September all of them had gone. It seemed strange to lose them but, of course, that was all part of the project and, like all good parents, the team had to learn to cut the apron strings. All the same, I found it much more difficult than my previous work with eagles and kites, which do not migrate and so can be followed throughout the year. With the migrant ospreys there was so much to worry about – would they head for their proper wintering grounds, would they survive the journey and would they come back home to Rutland Water to breed? In 1999 the first young chicks were fitted with satellite transmitters and we were able to follow them, sometimes watching them complete successful journeys to Senegal or Mali but, on other occasions, having to cope with the knowledge that they had been blown out to sea and lost in the ocean. It was a lesson for us all on the perils of migration. The information generated by the transmitters was regularly placed on the osprey website set up by Barry Galpin and followed avidly by many people, both at home and abroad.

In the days before transmitters, though, we could only wait and speculate about when or whether the first chicks would return. At home in Scotland, going about my work, I found that I was always hoping for that first phone call from Tim or Helen and it did eventually come on 29 May 1999. Chick number 8, released in 1997, was back. He was a young male and was joined two weeks later by another. The two young males stayed around all summer visiting the artificial nests, flying around Lax Hill

Helen Dixon at the cages, Rutland Water

and, later in the summer, joined the young ospreys released that year. They returned in 2000 and another two young males also returned, but it was not until 2001 that one of the first young males attracted a mate.

Breeding at Rutland Water

This male had built his own big nest in a tree near the reservoir, and it was here that he paired up with an unringed passing female, which laid three eggs in April 2001. The Rutland Project team was ecstatic and a group of staff and volunteers watched over the new nest and kept a detailed log of events. In June they were sure that a chick had hatched and I was thrilled to accompany them to the nest on 8 June so that we could check the contents with my mirror pole. We found one chick and two eggs that had not yet hatched. Tim later ringed the youngster, which fledged on 30 July and migrated on 20 August, two days after its mother and four days before its father. This was a truly momentous event: the first young osprey to be bred in the southern half of England for at least 150 years.

There were three other males present in 2001 but they did not attract mates, although three females stopped for short periods. The following spring, the breeding pair returned and laid three eggs. One chick hatched but, sadly, died soon afterwards during atrocious weather. It was a particular blow, especially since the other males had also failed to attract mates. In 2003, though, matters improved and the old male attracted a new young mate, a female that had been translocated in 2000. She laid three eggs that all hatched and the young were reared successfully. Another male, which had built his own nest in 2002, paired with a two-year-old female from 2001 and, surprisingly, this very young mother successfully reared two chicks. It was our first known case of an osprey breeding at two years of age.

We suffered a reversal of fortunes in 2004, but that is how these projects go. The young female failed to return and the old pair had only one youngster. Several other females were attracted by the unattached males and at one stage it really looked as though a pair would breed opposite Tim's house, but the female partner eventually left without laying eggs. But exciting news came to us during the summer of 2004 when it was discovered that males which had been translocated and released at Rutland Water in 1997 and 1998 had been identified as breeding in Wales. One of them had reared a single chick but the other nest had blown down, killing the two chicks in it. We also knew that one female released at Rutland

First successful nest

7th June 2001. Rutland Water. Got there at 3 p.m. nice day, rain showers. To Tim's cottage and met Helen, then to the centre and met Ian Newton and we walked through the woods to the new nest – to small hide. Watched male eating fish at side of nest, female incubating low down – then male away and female fed young from side of nest – well down in the cup. Then incubating again and male on low perch. Really good views across the field to the nest in dead oak tree. Next day with Tim, Ian and Helen to the nest by Land Rover; drove to the base of the tree and put up the mirror pole. 50 feet; both adults present – noisy but not coming low down – with mirror could see one young in the nest plus two eggs – drove away and female straight back to the nest. First breeding – great news. All very chuffed – back to Tim's cottage.

Pair at nest with chick, Rutland Water

Water had returned to Scotland and bred for three years near Dundee, rearing a total of nine young.

In 2005 the old pair returned and reared three young; while three other males displayed at nests and another was briefly seen. But two of the 'regular males' failed to return that year. At least five females were identified. One young female arrived on 28 April and stayed to 14 September. She spent time with several of the males while another late-arriving female was present at a nest from 3 August for six weeks. In view of the shortage of territorial males, I collected a new batch of eleven young ospreys (nine of them females) from Scotland and they were reared and released by Tim Mackrill and the project team at Rutland Water. Interestingly, the female that was there at the time fed these youngsters once they had been released.

In 2006 the old pair returned and reared three young, but the other males failed to attract mates to stop and breed. There was great excitement when both chicks from 2004, returned as two year olds. The female of the two settled down at a nest in Manton Bay, so there was great hope for next year. In 2007 the old pair reared three young again; they are experienced breeders, while the young female reared two young in her first year with the old male, which was the first osprey to return in 1999. So at ten years old he has at last got a mate as well as a nest and has reared two chicks.

The Rutland Water Osprey Project was the first of its kind in Europe and has been followed with close interest by raptor specialists in other countries, who would like to see ospreys restored to their own areas. (See Chapter 11 for projects in Spain and Italy.) Meanwhile, we are looking forward to seeing how the population will grow in the future. How will the birds fare in Wales and will others settle elsewhere? Our project has not just influenced Rutland but has changed the UK map for ospreys and should, we believe, lead to recolonisation in new areas. It is encouraging that the reintroduction experiment has worked but it has surprised me that so few passing females have stopped to breed. It is important to try and learn from the experience and change what we do in future. With hindsight, we should have taken more young females from Scotland to Rutland rather than taking a majority of young males, as was the practice in America. We should move more young females to Rutland and other ospreys of both sexes to new areas in the British Isles so that the bird can return to all its previous range.

Fourteen of the youngsters at Rutland Water were satellite tracked on their migrations and full details can be found on the Project website at www.ospreys.org.uk. There were four in 1999, six in 2000 and four in 2001: eight of them were females. A brief summary of each bird is as follows:

1999 R03: First to Lands End, then to France and migrated to River Niger in Mali.
1999. R04: Flew to Senegal, 4878 kilometres in 21 days.
1999. R05: Migrated to Senegal and then moved 233 kilometres north for the winter.
1999. R06: Radio stopped working in Kent.
2000. T01: Migrated over the widest part of the Sahara desert to Mali, 2100 kilometres in six days. Then had a week long tour of Guinea, Mali, Mauritania and finally wintered in coastal Senegal.
2000. T02: First went to Wales, then back to England and so to Morocco.
2000. T03: Flew Lands End to north Spain direct, flew later to Senegal.
2000. T04: Long stay in La Rochelle, France.
2000. T08: Wintered in Guinea-Bissau and Sierra Leone, crossed 2000-metre-high pass in Atlas mountains.
2000. T09: Died at sea off north-west Spain; direct flight from southern England, 1200 kilometres in 30 hours.
2001. U03: Radio stopped working in central Sahara.
2001. U04: Left England 23 August, off northwest Spain 27 August–2 September; later found dead at Puerteventura, Canary Islands on 8 September. An incredible journey for an osprey.
2001. U06: Wintered in Portugal, returned to England in summer 2002 and bred at Rutland Water.
2001. U10: Eastern Atlas mountains by 8 September; the radio then stopped working.

Breeding success at Rutland Water

Rutland Water	Pairs	Pairs with eggs	Young	Running total of young
2001	1	1 (3 eggs)	1	1
2002	1	1 (3 eggs)	0	1
2003	2	2 (3 eggs + 2 eggs)	2 + 3	6
2004	1	1 (3 eggs)	1	7
2005	1	1 (3 eggs)	3	10
2006	2	1 (3 eggs) + NB†	3	13
2007	2	2 (3 eggs + 2 eggs)	3 + 2	18

†NB – non breeding

The Lake District

In 1996 the first pair of ospreys bred south of the Glasgow to Edinburgh boundary, in the Scottish Borders and, of course, the population elsewhere in Scotland was continuing to rise. It was also the first year of the translocation of young to Rutland Water. There was always a chance that birds might naturally continue the southward spread and return to England, hopes that were boosted in 1997 when two ospreys summered in the Lake District. The following spring it was noted that the female at Bassenthwaite Lake carried a colour ring. It was a chick I had ringed near Kingussie on the 19 July 1995 so we knew that she was then three years old. That year, osprey enthusiasts in Cumbria built an artificial nest beside Lake Bassenthwaite in an effort to encourage her to stay. Young ospreys had by that time been released at Rutland Water and ospreys migrating through England could have seen these young birds and their 'distribution map' would have been changed.

In 1999 it turned out that a pair had attempted to breed at a secret site elsewhere in the Lake District but had failed; no one knew if they had laid eggs or whether they had just built

Breeding success in the Lake District

Lake District	Pairs	Pairs with eggs	Young	Running total of young
1999	1	yes	failed	non-breeding
2000	3	1	1	1
2001	3	2	3 + 1	5
2002	1	1 (3 eggs)	2	7
2003	1	1 (3 eggs)	1	8
2004	1	1 (3 eggs)	1	9
2005	1	1 (3 eggs)	2	11
2006	1	1 (3 eggs)	3	14
2007	1	1 (3 eggs)	3	17

a nest. Experience in Scotland would suggest that it was a young pair of birds, probably both three years old, that built a nest but did not lay. It was interesting that this was the summer when a young Norwegian osprey carrying a satellite transmitter, which I had caught in Scotland, spent time in the Lakes from 14 to 21 September.

In 2000 the pair of birds at the secret nest returned and reared one chick. This was the first successfully breeding pair of ospreys, not only in Cumbria but in England as a whole, for at least 150 years. At Bassenthwaite Lake, the first bird returned on 13 April and the second on the 17th but, although there was much nest building activity, mating and display, no eggs were laid and it was thought that the male was a young bird. A third pair built a new nest in Cumbria but they either failed or did not lay at all, being a young non-breeding pair.

In 2001 the original Cumbrian pair, amid great secrecy, reared two young while the Bassenthwaite pair, which was now on view to the public, reared one. The third pair either did not breed or failed. Bassenthwaite is a beautiful lake in a majestic setting between high mountains. The lake is several miles long and a mile across with a river, reeds and marshes. The surrounding hillsides are rugged and covered with conifer plantations and native woodlands, with a small farm on the shoreline.

That same year, I was very fortunate to be invited to the ringing of the first chick at Bassenthwaite. I stayed overnight with Peter Davies, a National Park officer, and at dawn we were at the Forestry Commission office to meet the team. We drove into the forest and walked through the wood to the Scots pine which housed the nest built by the staff of the Forestry Commission and the National Park to encourage ospreys to breed. The climber went up the tree and brought the single chick down to the ground where this first young osprey was admired by the team. It was ringed with a red colour ring, 'white 15' before being returned to its nest. A fresh, partly eaten roach was seen in the nest. Later in the morning, I admired the female bird with her chick standing on the nest, watching them from the public viewpoint. Ospreys last nested in the area in the early part of the nineteenth century and it was so wonderful to know that at long last, they were back!

Sadly, in 2002, it was back to one pair in the Lake District, and that was the Bassenthwaite pair, which reared two young, followed by only one youngster in 2003 – another chick died. At this time it was confirmed that the breeding female was the colour-ringed bird from Scotland, now eight years old. In 2004 the pair reared one young, from three eggs laid. An examination of the failed eggs revealed that one contained a half grown chick and the other was infertile. In 2005 a four-year-old osprey which had been translocated from Scotland and released at Rutland

Water in 2001 joined the female at the nest, 4–7 April, until the resident male arrived. The female laid three eggs and two young were reared. Next year they reared three young. The old female failed to return from migration in April 2007, but fortunately a new female arrived on 15 April, two days before the delayed old male. Three eggs were laid and a full brood of young reared. Another successful year to coincide with the half-millionth visitor to the Bassenthwaite ospreys.

It is disappointing that the other pairs stopped breeding early in the recovery, and it is a pity that the main pair did not produce three young every year, but soon some of the earlier youngsters should start to return. It is encouraging to note that there have been intruding ospreys around most summers and that this year the lost female was replaced immediately. Let us hope that more pairs start breeding because Cumbria is a great place for them, with plenty of fish to feed on.

There are many other suitable areas for ospreys to nest in England and, nowadays, there are many places where local people have built artificial nests to encourage them, stretching all the way from Devon to Northumberland. More birds are spending time at these sites during the spring and summer and it is hoped that the initial recolonisation of England taking place in the Lake District and at Rutland Water will continue. There were, in fact, reports of another pair breeding in the north of England in 2004.

Wales

When I worked for the RSPB during the 1970s, one of my good friends and colleagues was Roger Lovegrove, the Society's regional officer for Wales. I enjoyed watching red kites when I visited him and we sometimes talked about the possibility of ospreys breeding in Wales. To me, there were plenty of suitable habitats both for nesting and for fishing. Later, in the 1990s, I visited him for a short field trip with an ecologist from the Seven Trent Water company, which resulted in a couple of artificial nests being built for ospreys near one of their reservoirs. Like so many others before, though, they were never used. In 1998 and 1999, there was a summering osprey in mid Wales, a female which had been ringed on 29 June 1996 at Lake Muritz in eastern Germany, so at least it was suitable osprey habitat.

A few years later, the Forestry Commission asked Roger and me to look at the possibility of carrying out a Rutland Water style translocation of young ospreys to their forests in Wales. Between us, we carried out some fieldwork and wrote a feasibility study that set out why the proposal was a sensible and workable one. Despite the lack of written proof of ospreys nesting in the country, there was no doubt that this species used to be present, as proved by many cultural and historical references. There were plenty of suitable habitats to support even as many as a hundred pairs of ospreys in Wales. It was planned to start the project in 2003 or 2004 but the Countryside Council for Wales and the RSPB were not supportive of the proposal, and the necessary permissions were not obtained.

So, the discovery in 2004 that two pairs of ospreys were nesting in Wales took everyone completely by surprise. A pair of ospreys had been seen at the first site in the summer of 2003 and a new artificial nest had been built for them during the winter. They did not use it but a visiting birdwatcher later discovered them nesting in a Douglas fir. A small group, including Roger and the Montgomery Wildlife Trust, helped by the farmer who owned the land, kept a close eye on the pair. It was very encouraging that they hatched eggs and subsequently reared a single youngster. This was the first young osprey to fly in Wales for well over 200 years. It was also most revealing that once the birds had been carefully examined with a high-powered telescope,

Seeing other's efforts to attract breeding ospreys

2nd May 1996. Bough Beech reservoir, Kent. Time to spare at Gatwick airport en route from the States so to Bough Beech to meet guys involved with reserve. Looked at the osprey platforms built by Roy Coles, discussed ideas and what was needed to make them better. Some good Douglas firs as nest sites in the back. Two whimbrel flew by and a little ringed plover on the causeway, where there are osprey information signs. Looks good place for them to fish and to breed. Back to Gatwick.

it was discovered that the male had been translocated and released at Rutland Water in 1997. The female was colour ringed and she later proved to have been ringed by Keith Brockie in 2000 in Perthshire. Disappointingly, the pair did not return to the eyrie in 2005.

The second Welsh pair was found nesting near Porthmadog in North Wales. They also chose to nest in a tall Douglas fir and, because they were in a more public area, the RSPB decided to make the news public and man a viewpoint with partners from the Countryside Council for Wales and local people. Sadly, during a day of strong winds, the nest fell out of the tree and the two young died on the ground. In the end, it was possible to check both parents closely with a telescope and, although the female was un-ringed, the male turned out to be another Rutland Water bird that had been translocated and released in 1998. I visited the nest area in winter and was asked for advice on how best to build a secure artificial nest in the same tree, and this was built by the RSPB in the winter. The pair returned to rear two young in 2005, and the same pair bred in the last two years, rearing two out of three young in 2006 and two young in 2007. Large numbers of people have visited the public observation post to watch the ospreys at their nest. On a few occasions an extra osprey has also seen in the vicinity.

For the Rutland Water project team, this discovery was really exciting news. We always knew that some of the released birds would not necessarily return to Rutland Water but it was interesting that they had returned to the same general latitude, and that some of the young ospreys on their first migration had initially moved west. We now

Looking for artificial osprey sites in Wales

3rd December 1997. Very cold but sunny all day. Roger and I left his own home early and drove to the Severn-Trent reservoir to meet Dee Doody; drove round the reservoir and found three places suitable for artificial osprey nests – on the hillside in stunted larches. Dee to build them later. Four red kites flying about and also ravens. Then Dave from Severn-Trent Water arrived and we went past the pumping station and round the reservoir discussing the project to encourage ospreys. Had a pub lunch at Rhyader and then called at Gingrin farm to watch the kites feeding – about 20 came in. A fine big Welsh Black bull. So to a Welsh Water reservoir where very good tall Douglas fir trees suitable for ospreys and also a possible artificial site: a Scots pine tree on an island. So back to Roger's at Newtown.

knew that the release of birds at Rutland Water was indeed changing the pattern of recovery in the British Isles.

In some ways it was amazing to us that the Rutland Water project should have changed the face of the osprey population in Wales, even before it had been possible to carry out the translocation project on Forestry Commission land. There is a very strong case that these pioneering pairs, isolated from the main population of ospreys, should be bolstered by the translocation and release of more young so that a strong population can be established in a new part of the British Isles. These projects are so influenced by chance. What if the German bird had arrived one summer later and met the Rutland Water male? There might then have been breeding ospreys in Wales in the year 2000, which would have made a very cheering start to the new century.

Breeding success in Wales

Wales	Pairs	Pairs with eggs	Young	Running total of young
2003	1?			non-breeding
2004	2	2	1 (mid Wales)	1
2005	1	1 (3 eggs)	2 (north Wales)	3
2006	1	1 (3 eggs)	2 (north Wales)	5
2007	1	1 (3 eggs)	2 (north Wales)	7 young

In 2003 I attempted a map-based estimate of the potential population in the British Isles, using our present knowledge, and accepting that ospreys will continue to change their behaviour. This estimate suggested potential populations of 500 pairs in Scotland, 700 pairs in England, 100 pairs in Wales and 400 pairs in Ireland. This would represent nearly a 90% increase in numbers, which indicates the restricted nature of the present range and population of the osprey in Britain.

Eyrie in Wales, 2004

The first breeders in Wales for 200 years!
26th July 2004. Mid Wales. Met Graham Williams, Tony Cross, and Clive from the Montgomery Trust – coffee in the café. So to the nest, female on tree beside nest and could see red ring but no sign of young – then male came in later with a small fish. By this time the hired 'cherry picker' had arrived and Tony Cross drove it through the fields to the nest. Unfortunately the cherry picker was not high enough because we could not get it close to the base of the tree due to the hedge. One half grown young in nest and male and female noisy overhead. I found remains of last year's nest in live Douglas fir next to dead nest tree. Checked ring numbers with telescope from the gate – male has 'white-black 07' on left leg and female has 'red-white 6J' not 8J on left leg. The first successful nest in Wales.

11
A world of ospreys

The extraordinary range and adaptability of the osprey allows us to encounter it in nearly every corner of the world, against amazingly varied backdrops. I have seen ospreys at both ends of their heroic migration: in the pine forests of Scandinavia, and safely arrived at their African wintering home. I have watched ospreys in the salt marshes of New England, USA, on the coral islands of the Arabian Sea, on the great sea cliffs of Hokkaido, Japan, and nesting on steel pylons in northern Australia. And, wherever I have been, the people who study them as I do have added to my knowledge and my enjoyment of these unique birds.

Mainland Europe

There are close ties between our ospreys and those of mainland Europe. I still look back to the excitement of straining, but failing, to read the complete ring number on the male bird at our second nest in 1963. Telescopes have come a long way since then and the bird, in those days, was just too far away. I managed, though, to read two of the numbers and identify it as a Swedish ring, finding out later that the osprey had been ringed near Stockholm. And Sweden is the right place to start this journey round the world of ospreys, being home to Europe's largest population, and the place where most of the ospreys that we see on migration in Western Europe have originated.

Sweden

When I was young, I learnt most of what I knew about ospreys from the *Handbook of British Birds* and writings about the birds in Sweden. I particularly remember an article written by George Waterston in *British Birds* journal, accompanied by superb photographs of a pair of ospreys nesting on a Scots pine among the Swedish lakes. To me, it looked very much like those parts of the Highlands where ospreys nested in the 1960s and I so wanted to get there. My first visit to Scandinavia, however, was not until early 1972, and then it was to northern Lapland, where all I saw of the ospreys' breeding grounds were tantalising glimpses from our plane of the massive forests dotted with lakes, where, far below me, up to three thousand pairs were breeding.

I returned in 1979 on a month's sabbatical from my job with the RSPB, principally to look at fish farms and their relationships with wildlife, but also to understand more about ospreys and goldeneye ducks. After visiting Denmark, I crossed to Gothenburg by car ferry and drove north.

Sweden. 12th June 1979. Driving north near Lake Vänern, fantastic area of lakes with pine and spruce forests and small areas of farming – walked in several areas – lots of goldeneyes plus nest boxes everywhere: several pairs of cranes in forest boggy areas including a pair with two small young in a small field by the forest edge. Met some hunters who were also birding in this remote forest and spent some time with them looking at moose and beavers – at least five in the little river. They were very involved in wildlife management as well as hunting moose and putting up nest boxes. Shown an osprey's nest on a small Scots pine in open moss within the forest, where several greenshanks calling and several pairs of red-throated divers flying around calling. Female osprey was incubating low in the nest – I watched from a distance.

After that I set off to visit Sten Osterlof, director of the Bird Ringing Scheme, who had ringed more Swedish ospreys than anyone.

Sweden. 13th June 1979. Went to the Natural History Museum in Stockholm and spent the afternoon talking about birds. Evening with Sten to see an osprey site occupied by a breeding pair – very similar to our Scottish sites, in a tall Scots pine in rather broken hilly country with some fine lakes nearby. Redwings singing and a pair of Slavonian grebes. This was the nest where he had his accident. He climbed the tree on a very hot summer's day and blacked out at the top and fell all the way down to the ground and broke several of his limbs. He has promised his wife not to climb high trees any more.

Sten Osterlof, Stockholm 1979

Next morning I travelled north up the Baltic coast, then inland into the mountains of Jämtland and slowly back south to Värmland, before moving westwards to Oslo. I saw ospreys fishing in many places, and also a number of nests, all of them on the top of Scots pine trees. Some were on islands in big lakes and others were in small mosses or peat bogs within the forest. Many of them looked very similar to the ones I knew in Scotland. It was also very exciting to find them living beside birds which I occasionally saw as rare breeders in Scotland, species such as redwings, fieldfares, green and wood sandpipers as well as common cranes, which I would love to see back in Scotland.

In later years I stayed at Falsterbo one autumn to visit the famous bird migration site and bird observatory, and, in the years 1989–1993, I visited the southern province of Skanor to collect young red kites, under a special licence, for the successful reintroduction project to Scotland. There in southern Sweden I saw migratory ospreys. Among the ornithologists that helped our project was Nils Kejllen, who has carried out important raptor migration studies.

Falsterbo. 11th September 1983. Early morning migration watch at the end of the point along with several hundred birdwatchers. Lots of yellow wagtails, tree pipits, robins, redstarts and also a very good passage of raptors; 130 honey buzzards, 69 sparrowhawks, 2 hen harriers, 9 merlins, 5 hobbies, 56 common buzzards, 5 marsh harriers, 22 kestrels and during the morning a total of 6 ospreys. Single birds flying steadily south except on one occasion when two flying together.

In the 1990s I met and started to correspond with Dr Mikael Hake who was carrying out field research on ospreys from Grimso wildlife research station. He also gave me advice on the satellite tracking of ospreys, which he had started to work on with Nils Kjellen and Thomas Alerstram of Lund University. One autumn, after attending a conference on sea eagles on the archipelago near Stockholm, we spent several days exploring his study area and discussing ospreys.

Sweden. 18th September 2000. To see Mikael Hake's study area – redwings and fieldfares flying south. Walked to the first nest site through forests to an artificial nest in a low pine on the edge of a bog. Next to a used nest by small lakes and then walked to a nest in a very big moss – an eyrie in pine tree on a hill – hazel grouse further on. He said he had some problems with goshawks killing young ospreys and further north in Sweden, eagle owls have been killing both adult and young ospreys.

I later learnt that a new survey of ospreys in Sweden gave a population of between 3297 and 3592 pairs, and that the decline caused by effects of pesticides in the 1960s had been reversed.

Norway

Although I have been to Norway on many occasions, my osprey connection with the country involves Norwegian ringed individuals which have started to breed in or visit the Highlands. These birds are from south-east Norway, near Oslo, and the total Norwegian population is between 210 and 260 pairs. This population is closely linked to neighbouring Sweden, but it now also has a close link with Scotland.

Finland

About 1200 pairs of ospreys live in Finland, and although I have only ever had a brief visit to the far north of Finland while bird-watching in Norway, I have had a long friendship with Pertti Saurola, who has studied Finland's ospreys over many decades. We first met at an osprey conference in North America in 1980 and at Loch Garten. We have corresponded and met at various osprey meetings, so that Pertti's information on ospreys in his country has always been readily available and eagerly read. I will visit one day; and often think of his satellite-tracked male, Harri, which bred north of the Arctic Circle and wintered near Cape Town – an incredible journey of 11,552 kilometres.

Germany

I first read about German ospreys in a very early book on the species written by Karl-Heinz Moll, an ornithologist in East Germany. One of the things that interested me most was the number of ospreys nesting on power lines. During the Communist era it seemed impossible that I would ever get to see those birds, but a chance meeting in Mongolia in 1980 with a group of East German ornithologists resulted in an invitation to come and see wildlife in their country. Getting the visa to enter East Germany was a tortuous process but, in the end, I got there.

East Germany. 22nd May 1984. Drove over the Elbe River and to West German customs – all very blasé – then to no man's land – open area – big fences and ploughed land along side and so into East Germany half a mile away. Ninety long minutes getting through customs post – golden orioles in song all around but I was scared to look at them while being processed. Finally away to Ludwigshurst and to my designated campsite at the Malcher See (Lake). Werner Eichstadt (who I met in 1980 in Mongolia) arrived with a friend and took me birding to Lake Waren – saw my first pair of ospreys nesting on top of an electricity power pylon (a square topped pylon) and then two more nests on pylons in a line, one with a female incubating but no sign of the male. Next day saw more ospreys as well as breeding sea eagles including a huge nest in an isolated birch tree, black stork, white storks, cranes, red and black kites, and into a beautiful beech forest where red-breasted flycatchers singing and a middle-spotted woodpecker. What a fantastic countryside and yet for me scary being on the other side of the Berlin Wall.

In the late 1980 a young German studying for a summer at St Andrews University came to speak to me about ospreys and I invited him to join me in my osprey fieldwork. Daniel Schmidt became fascinated by these magnificent raptors and started to study the species back home. In the early 1990s he became interested in whether ospreys could be restored to the Black Forest and his Institute asked me to come and give advice. I was amazed at how few large trees were suitable for ospreys to build natural eyries: the forests were simply too well managed for them and broken tree tops were a rarity. Fish supplies, though, were excellent and ospreys could be restored using artificial nests.

Daniel's ideas for southern Germany did not get support from his colleagues but later in the 1990s he started his PhD studies on ospreys in northern Germany. Several times I went to see him and some of his co-workers and their ospreys. It was a really exciting time for the bird as its population continued to climb, nowadays standing at over 500 pairs. Sixty to seventy per cent of nests are on electricity pylons, and many tree nests are also man-made.

Woblitz, Germany. 6th April 2001. With Daniel Schmidt checking Osprey nests on way to Lake Muritz – good number of birds back at nests in electricity pylon nests – using telescopes to read ring numbers – superb osprey nest in big old Scots pine on the grassland at a Galloway cattle farm – both adults busy nest building. Checked a couple of nests on pylons where Daniel had radio tracked birds and then onto a marshy area (Daniel into a hide so he could read the colour ring on osprey). Good number of cranes around. Later taken to see a black stork nest deep in the forest on artificial nest built by Paul Sommer – viewed with a telescope from 200 metres – one adult perched on the nest and then we saw the other's head just poking over the sticks as incubating – brilliant.

France

The first ospreys I visited in France were on the beautiful island of Corsica in 1982. I met and started to correspond with Jean-Claude Thibault, who worked at the Natural Park and had studied ospreys since 1973. He told me that in the early 1970s, the Corsican population was down to three pairs but following protection of the nesting sites and publicity by the National Park, the population rose to 10–12 pairs. The nests are all sited on the coast, principally in very high sea cliffs on the west coast of the island.

Some of the birds are sedentary while others, especially young, wander in the Mediterranean region; and they had just had their first recovery from North Africa. The breeding season lasts from March to August, and egg laying by different pairs is spread over two months.

Corsica. 10th September 1982. Long walk in the sun to a Genoese watchtower at the end of a big headland – very red granite with amazing shapes and pinnacles. Jays, Sardinian warblers, cirl buntings, and also 18 bee-eaters which were beautiful circling over the cliffs and calling all the time. Two crag martins flying around and from the top of the tower could look over a rampart and 1000 feet below to the blue sea. Three-quarters of the way down there was a big osprey nest, at least a metre high, on a rock pinnacle and another nest about 100 feet lower on another buttress. Both very fine nests and used alternately and I'm told they are guarded when in use by a warden who stays in the tower.

Above: Artificial nest built by National Park osprey wardens, Corsica
Below: Nest and osprey, Corsica (photos courtesy of J-C. Thibault)

Some of the nesting areas are quite similar to parts of the Scottish coastline and I was able to imagine places in the Highlands where ospreys could nest on coastal cliffs, such as the Black Isle and Easter Ross, if the population continued to rise and a few pioneering birds changed from tree to cliff nest sites. Maybe they did long ago.

The Corsican ospreys have continued to thrive over the years and the population reached a peak of 32 pairs, and is now all but full at 26 pairs. Many interesting studies have been made on the Corsican ospreys, particularly on breeding success and behaviour, colour ringing studies, colour patterns of individual birds and on innovative conservation efforts using artificial cliff nests and imitation ospreys to encourage them to colonise the north and east parts of the island. More recently, the Corsican Natural Park has provided some young ospreys for translocation to the coast of Italy.

Ospreys were first found breeding again in mainland France in 1984 and this population has also slowly increased until there are now 20–24 pairs resident in the watershed of the River Loire, a beautiful area of great forests and lakes, with single pairs in two other areas of France. The Loire population has been studied extensively for more than 20 years by Rolf Wahl. Most of the birds are colour-ringed, artificial nests have been built and, with assistance from Daniel Schmidt, individual adults have been captured and colour-ringed. In this way, at least three birds from the German population have been identified in the Orléans forest. I have visited the area on several occasions and helped with advice on colour ringing and artificial nest construction, as well as on possible translocation projects within France, for there is no doubt that there was, long ago, a very large population in that country. There could easily be once again, as France has such excellent nesting and feeding habitats throughout the country.

Orléans nest (photo courtesy of Rolf Wahl)

Orléans Forest, France. 3rd June 2000. Visit to the osprey breeding sites with Rolf Wahl; looking at the Loire River where his ospreys fish – breeding Mediterranean gulls, 12 whiskered terns flying up the river. So into main forest and checked another osprey nest, both birds present on Scots pine tree at beautiful nest site on a long lake. Checked several other ospreys' nests in the forest – all in big Scots pines and was told by Rolf about their histories. Also saw booted eagle and short-toed eagle nests. Previous evening had visited a bird hide overlooking a lake with an osprey nest on the other side – had discussions with the local people there about closed-circuit television possibilities. Nightjars on the road as we drove home in the dusk.

Spain

My first contact with ospreys in Spain was on the island of Minorca, where Rafael Triay invited me to come and give advice on conservation research into their small population of red kites, and also to see their breeding ospreys.

Minorca. 29th May 1993. Another beautiful hot sunny day. With Rafael and Giuseppe to Fontella. Launched rubber boat at Fontella – three very pale ospreys soaring over the village bay -- white houses with orange roofs, waterfront cafes and shops. One osprey – a male very white outer tail feathers and no dark breast band. Out in boat and to the west to look at the nesting sites of ospreys in tallish cliffs on north coast of Minorca. Most isolated site guarded at weekends. Across the bay to lighthouse where artificial nest built. Three young peregrines fluttering and chasing around the cliff top, juvenile shags – one-man fishing boats. Had lunch in the cliff top lair with guards,

scanning one of the osprey nests. Ringed female on nest with three-week-old young but no sign of male. Female hungry – she looked very much like one of ours with well marked breast. Nest on a ledge about 300 feet above sea level with great views.
Back down to the shore and found that the rubber boat's patches had lifted in the hot sun! Tried repair but not much use, boat not going well so Rafael dropped us off on isolated beach – sought shelter in caves as so hot. Kestrels three, booted eagle, peregrine, as well as swifts, swallows, blue rock thrush, stonechat and a quail. Three hours later Rafael arrived with car on a rough track – a long way back through deserted farms to the main road. Nice evening meal of anglerfish in almonds. Barn owls feeding young over the rooftops and stone curlews calling on the edge of town.

Rafael Triay has kept in touch ever since and I learnt that the small population on Minorca slowly rose to seven pairs. His studies using satellite radio transmitters proved that Mediterranean ospreys were not necessarily sedentary, as his young birds went as far as the western Sahara and up the River Rhône into mainland France. Sadly, this population, which nests on cliffs and feeds on sea fish, has started to decline in recent years, and worryingly there are now only four pairs left.

My first visit to mainland Spain specifically for osprey studies was in 1999, when I went to search for the breeding female from nest A11 near Carrbridge. She was well known to me and the first osprey that I ever fitted with a satellite transmitter. She migrated to Extremadura in Spain and surprised us by not travelling on to Africa. Her radio signals came from a very large reservoir, called Embalse Alcantara, in the Monfrague region – famous for all sorts of interesting birds. I decided to try to find her.

2nd November 1999. Midday driving down the eastern shore of Alcantara reservoir, when I saw an osprey several hundred feet above the water. Got the car off the road and watched, a small group of jackdaws hassling it. Using my telescope on the roof of the car I could see that it was a large dark female and suddenly, as she turned under the sun, I could see the tiny aerial of my transmitter sticking out from her back. Amazingly, it was our osprey from Scotland. She moved south into another bay and quite quickly caught a fish before heading west into the cork oak forests. Later, I checked the most frequently used location, from our satellite data, and found that it was a huge nest belonging to white storks in an isolated tree, which she must have used as a roost tree.

At this time, there were no ospreys nesting in mainland Spain. They had, in fact, been absent for nearly half a century. There was just one pair in the whole of Iberia and they nested in Portugal.

Our studies of osprey migration using satellite radios increased our interest in ospreys in Spain, because it is on their direct migration route. Birds often used that country for stop-overs before heading on to Africa, and there was some evidence that increasing numbers of northern ospreys were starting to winter in Spain. Of particular interest to us were the birds which made a crossing of the Bay of Biscay and arrived tired on the north coast of Spain. We could see from our research that some of them died there, but why? Following a visit to that coast, it was in my view due to high numbers of large gulls which would chase tired migrants and force them into the sea, before they could make landfall. This behaviour had already been noted with other tired raptors in the Straits of Gibraltar.

North Spain. 22nd September 2001. Passed Cabo Ortegal and so to the lighthouse – where ospreys must arrive from over the Bay of Biscay – lots of gulls in fact ca. 250 yellow-legged gulls on the rocks and jagged little islands – Corys' and Manx shearwaters offshore and migrants coming in – wheatear, willow warblers and a wryneck and then an osprey circling above woods to the west on top of the hill. Went round that way and found a high road with great views onto the estuary of the Mora River. Found an adult male Osprey, perched on a post in the estuary – got it to fly – no rings. Then the rain got worse!

My next interest in Spain was to learn more about the ospreys that wintered there and I was fortunate to meet Jose Sayago and his colleagues at the Odiel National Park. In 2002, he had identified 33 wintering ospreys along this part of the coast between Huelva and the Portuguese border, and he had learnt how to catch them. In this way, he had identified three Scottish, two German, one French and one Finnish ringed ospreys; one of the Scottish birds was eleven years old.

The author with Jose Sayago and colleagues at the Odiel National Park – with Scottish osprey

Odiel marshes, Spain. 8th March 2002. Went to the Isla Cristina marshes and at 1110 hours Sayago and Enrique left me in a hide out on the salt marshes, lying down about 30 metres from their osprey trap. 1120 hours my radio went quiet after the guys called from further away. Very still, but not too hot in the hide. 1142 hours 'Roy – Roy – osprey coming' came over my radio. Almost immediately I could hear the beat of its wings above me and suddenly it landed on the trap and was instantly caught. Jumped out of hide and photographed it hanging there; then caught it and amazed to find that it was a Scottish ringed bird. Collected by the wardens and taken to the shore. Large female. Ring number 1351697 weighed 2200 grams – very heavy!; wing length 520 millimetres and wingspan 1.71 metres; he added a yellow colour ring. Bird released and flew off over the marshes. This bird had arrived in the marshes in 1999 and had returned each winter to same area. Ringed as a chick in Scottish Borders by Davie Anderson.

Left to right: Rolf Wahl and Jose Sayago at Odiel, 2000

Male osprey in Spain, March 2002

Earlier in my field trip I had visited the coast to the south of Seville and watched ospreys in several parts, including the estuary of the river Guadalquivir and the salt pans and fish farms. During this visit I had discussions about osprey translocations with Miguel Ferrer and Eva Casado of the Doñana Biological Research Institute, following our Rutland Water Project. They were keen to restore the osprey to mainland Spain and so developed a re-introduction project in partnership with Scotland, Finland and Germany. I had already seen one of the potential re-introduction sites in 2001.

A German osprey caught for radio tracking in winter, Barbate reservoir

Barbate reservoir. 30th September, 2001. Manolo and Ramon walked us in to the edge of the reservoir on a side arm where an osprey nest had been found, built in a dying eucalyptus tree. A very good place for ospreys to breed as long as they could deal with the white stork competition for nests.

Barbate reservoir. 30th March 2003. Went to usual area and now 10 pairs of white storks in the dead trees. An osprey adult feeding on fish on old electricity pylon. Back to the highway and out to a new search area with local farmer and found an excellent release area for hacking cages on opposite headland. Adult female Osprey flew by – lots of mallards and three garganey. Excellent slope down to water and shaded by trees with an old house. Great place to build the hacking cages for releasing young ospreys.

In July 2003 the first four young ospreys from Finland were released at the site and the following summer young ospreys from Scotland and Germany were released there, and additional young from Germany and Finland were released from hacking cages at Odiel marshes. More have been released each summer since.

30th July 2004. Up at 3.30 a.m. – fed young ospreys and Rona, my daughter, and I got them ready to travel. On to Inverness to the cargo department at airport – sorted transit details to Seville today. BBC turned up and also Mike Scott (Deputy Chairman SNH) – did interviews. Birds on first flight to Gatwick and then we all flew on to Seville on time as well. Met Eva at the Customs and also Spanish TV; soon through and got the birds from cargo and then loaded into their 4 x 4 with Matias – quickly travelled down to Barbate. Put the birds in the release cages, which are excellent and also checked out the six German birds which were all looking good.

Camera crews await the arrival of the translocated ospreys to Spain

Left: Young Scottish osprey en route by air to Spain

Below: Miguel Ferrer and Eva Casado ringing osprey chick at first Spanish nest

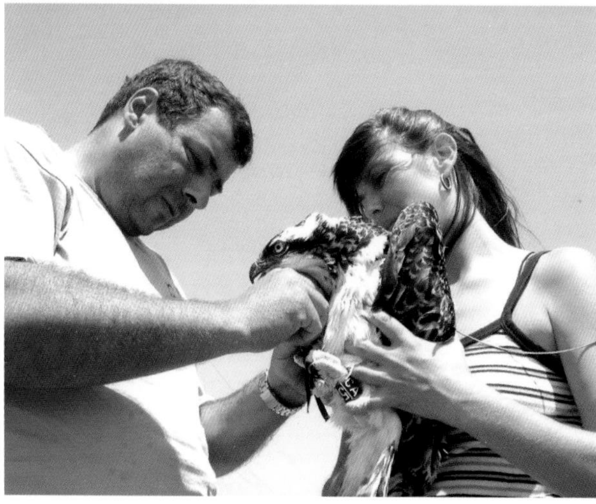

At our annual workshop to discuss the re-introduction project in March 2005, not only were we told about the success of the releases the previous autumn, but Miguel and Eva had exciting news about a pair of breeding ospreys. They were not our young birds but it was believed that they had been stimulated into breeding in this area because of their interactions with the ten young ospreys released the previous autumn.

16th March 2005. Pair of ospreys at a nest on a small square pylon in northern arm of reservoir – drove to farm on private road – one bird looked as though incubating and other on edge of nest – little later they changed over, male away to fish on the long arm and back in five minutes with a fish and ate it on the shingle spit. Female low down in nest obviously just laid egg – as not incubating yesterday. Great news – first breeding Mainland Spain for many years – so to local village for lunch and to celebrate!

Portugal

There used to be a population of breeding ospreys along the coast of Portugal, but in the last century numbers declined drastically, due to shooting and disturbance, leaving just one pair. In the late 1990s I was asked by a group of Portuguese ornithologists about the possibility of reintroducing ospreys. Pedro Beja and Rita Alcazar came to see our project at Rutland Water and the breeding ospreys in Scotland. In the autumn of 1997 Luis Palma from Faro University and his colleagues organised a workshop of experts to discuss their project, which they then put to their government in Lisbon. Unfortunately, they never had a positive reply and, finally, the last male bird disappeared as well. Ospreys are now extinct in Portugal.

Alentajo coast, Portugal. 27th September 1997. With Luis, Pedro and Rita looking at the potential of restoring ospreys to Portugal. Went to several places, looking at cliffs and old and potential nesting sites – some superb coastal scenery – next on south to a superb coastal site at Arrifano where the last nest is on a dramatic rock pinnacle out in the sea, male was present during summer but female died in the spring of this year – they were the last pair in Portugal. The military winched down a man from a helicopter to retrieve the dead bird which was entangled in fishing net. Looked at nests on the mainland cliffs but many were in fact white stork nests. Plenty of grey mullets in the estuaries.

Netherlands

In 2002 a pair of ospreys built a nest in a small tree at a nature reserve in the Netherlands, but they did not lay eggs. Ospreys had not nested in the country for over 100 years.

Oostvardensplassen. 24th March 2004. Asked to give a workshop on ospreys by the State Forest service. And to discuss artificial nest building following the presence of a pair of ospreys which built a nest in 2002 but they did not use it in 2003. Next day we built two really good nests in the willows and also found at least one other really good nest site, although large willows are not particularly good nesting trees because they rot so quickly. This is a really good place for ospreys with plenty of food and would be a good place for a re-introduction project. Also saw great white egrets and the sea eagle as well as the feral Hek cattle and Konik ponies.

Italy

Although ospreys migrate through Italy, none breed in the country: they were persecuted to extinction long ago.

La Maremma National Park, Tuscany. 30th March 2004. Invited to give a talk on osprey reintroduction to Italy and also field visits to find suitable sites on the National Park for a translocation from Corsica. Found an excellent place for release cages and also looked at the artificial nest they had already built. Just as I was leaving on the first of April I saw male osprey flying up the river and perch in a big poplar on the riverbank and then about 10 minutes later another flying up the river about half a mile further on – so obviously a good place to see the species. Saw the local Maremma cattle as well as four wild boar and some red deer, also excellent views of a great spotted cuckoo.

In 2006 Andrea Sforzi and his colleagues at La Maremma National Park started their re-introduction project and seven young birds were brought across from Corsica and reared and released in Tuscany, with another six in 2007. Let's hope they are successful and a new population of ospreys is established in Italy.

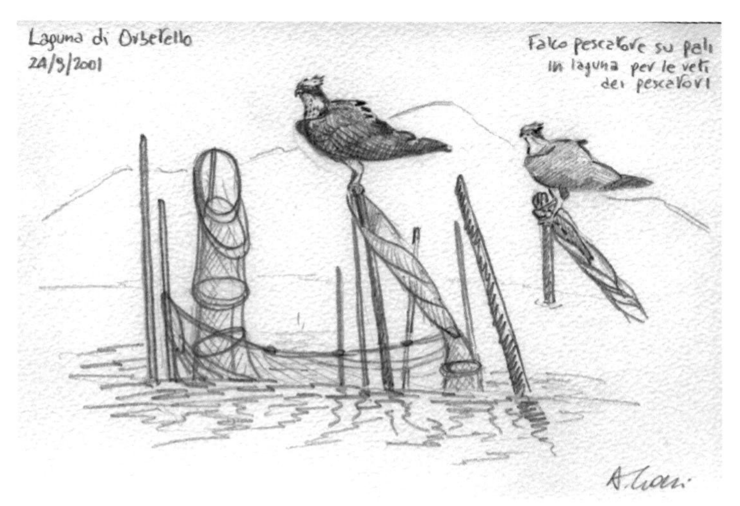

Sketch of osprey in Italy

The recovery of the osprey in Europe

The osprey is a pan-European breeding species that has suffered intense persecution and consequent reduction of its historical range. It is missing from much of southern and western Europe and although northern and some central European populations are generally healthy or in some cases expanding, there is little or no evidence of major range recovery elsewhere. Following the examples of North America and Rutland Water it is an appropriate species for reintroduction projects.

Ospreys are legally protected throughout Europe and, as the decades pass, people have become prepared to accept them within the landscape, even in highly populated areas. Once established in new areas, ospreys are often carefully protected by the local human population and breed successfully. Consequently, they become more tolerant of humans and can increasingly accept people closer to their nests and feeding areas. They are also capable of nesting on man-made structures.

The European Habitats Directive encourages member states to examine the opportunities for active restoration of regionally extinct species. Range recovery would be highly beneficial to European ospreys and a larger and more widespread population would reduce their vulnerability to changes in climate, pollution and variations in food supply. The potential size of the population is speculative but I made an attempt, assuming a programme of protection of pioneering pairs, artificial nest building, translocations and reintroductions throughout the countries where the species has been lost. I then examined the potential for Denmark, Germany, the Netherlands, Belgium, France, Spain, Portugal, Switzerland, Austria, the Czech Republic, Slovakia, Hungary, Italy, the Balkans, Romania, Bulgaria and Greece. My estimate was of the order of an additional 5000–6000 pairs of ospreys throughout these countries.

After probably half a millennium of persecution the ospreys of Europe reached their lowest point in the middle of the last century, but in the last fifty years there has been an upswing in the population in many countries caused by greater protection and a greater understanding of conservation. Most populations in Europe are increasing, although the situation with the large population in Russia is less clear and they may have declined. Nevertheless, the bird is still very rare in the southern half of its range in Europe and the European population is probably at half its true potential, presuming it successfully reclaimed its lost range.

The latest country to lose breeding ospreys was Portugal in 2002; ospreys are also missing in Ireland, the Netherlands, Belgium, mainland Spain (although there was one new pair in 2005), Switzerland (1911), Austria (1930s), the Czech Republic, Slovakia, Hungary, the Balkans, Romania, Greece (1966), Turkey, and Italy–Sicily (1955).

Numbers of breeding pairs of ospreys in present day Europe

(The data is mainly compiled from an osprey workshop undertaken in Spain in 2005 and more recent data)

Country	Number of pairs	Trends
Scotland	c.200	increasing
England	3	new since 2000
Wales	2	new since 2004
Russia	2000–4000	stable or decreasing
Ukraine	5–10	
Sweden	3297–3592	
Finland	1200	stable or small decrease
Norway	210–260	increasing
Denmark	1–3	
Byelorussia	120–180	
Latvia	120–150	increasing
Estonia	40–45	increasing
Lithuania	c.50	
Poland	70–75	stable or decreased
Germany	500	increasing rapidly
France (mainland)	25	
Corsica	30-26	decreasing
Majorca–Minorca	21–15	stable or decreasing
Algeria	9–15	probably decreased
Morocco	20	probably decreased
Ukraine	5–10	
Armenia	1–4	
Azerbaijan	0–5	
Moldavia	0–2	
Bulgaria	0–10	
Canary Islands	16–21	
Total approximately	7945–10423	

Africa

Nearly all the ospreys which breed in northern Europe, including western Russia, winter in Africa. Our Scottish birds go to West Africa: arriving in September, they spend half their year in the lakes, marshes and rivers there, as well as along the Atlantic coasts from Mauritania to Guinea and inland to Mali. The more easterly breeders, such as in Finland, are more likely to be found in Nigeria or east Africa and sometimes as far south as South Africa. Africa is very important for the conservation of European breeding ospreys and I hope there will in the future be better partnerships between the two continents for the conservation of migratory raptors.

Various osprey researchers have been to Africa to study them and I was fortunate enough to visit West Africa in 1977 when the RSPB was making a film about ospreys. In other years I visited Zimbabwe, Tanzania and Kenya – where I saw single ospreys – and more recently the Cape Verde islands off West Africa, where there is an extremely interesting population of breeding ospreys.

Sketch of osprey in Africa

The Gambia

In January 1977 I went to Gambia for a month with Hugh Miles to film ospreys in the winter as part of a major RSPB film. Hugh filmed ospreys on the coast, sometimes from temporary photographic hides, and from boats in the mangrove swamps and on the main river. While he filmed, I carried out studies of wintering ospreys. We often saw the birds fishing well out to sea, and even returning with flying fish. The coastal areas of Senegal and Gambia are very important for our birds, and while there was plenty of fish at that time, it is very worrying that massive industrial scale fishing carried out by European fishing boats is already damaging the livelihoods of local fishermen and probably the lives of fish-eating birds as well.

The local catch, Gambia – up until now the fish has been plentiful and varied, but industrial fishing is taking its toll on the local fishermen, communities and ecology of the area, as well as having consequences for the osprey

Sunday, 9th January 1977. Arrived at Houba beach at 10 a.m.; light onshore wind and seas slightly choppy. Two to three ospreys fishing in the bay; one caught fish immediately and set off inland behind the South fishing village. 1130 two more in Bay plus a black kite and palmnut vultures. 1200 one caught fish half mile offshore and headed low inland over our heads with one and half pound flying fish; wings of fish very obvious in flight. 1215 five ospreys in the bay, some over a mile offshore. 10 black kite trying to lift something out of water. 1230 osprey caught flying fish off the

Mangrove perch in the Gambia

Saniend point and flew across the bay at 100 ft chased by Caspian tern and briefly by a black kite – a very persistent tern chased osprey right across bay for over a mile and even caught hold of the fish several times by making attacks from behind and below. A very interesting attack method. Osprey went inland to the palm trees. 1405 another osprey caught a flying fish by tail end and had hard struggle flying against wind with the fish spreading its wings in the wind, it struggled along past Saniend point and the boats, and had to land on a rocky point to rest where it was mobbed by grey-headed gulls. But no real attempt to get the fish from osprey.

23rd January 1977. To Jean's house at 5 a.m. for Land Rover trip of 115 miles inland to Mansa Konko. Most beautiful dawn glow through the savannah – a few owls and a long-tailed nightjar on the road. Lovely dawn breaking as we got to the ferry car park to unload Zodiac rubber boat then off down river. Jean Eschenlour, Hugh and myself plus all the gear. The river Gambia is immense and one feels like being on a sea rather than a river – mangroves on both sides up to 100 feet high merge together and so the bends disappear. Super views of an African fish eagle flying along above the river and then fluttered down and grabbed a fish with one foot and didn't get wet. Ten knob-billed geese, 300 pelicans, 30 wood ibis, white-winged, black and gull-billed terns. Turned in to a side creek called Jurang Bolon – a completely different world and absolutely beautiful – meandering creek about 50 yards wide fringed by mangroves up to 100 ft high and mud banks. Saw two crocodiles – big ones – hauled out on the bank but they dived straight back down into the water before they could be filmed – also marks on the bank of several others. One otter in tree roots also fish eagle and pied and blue breasted kingfisher, three finfoots and two bishop storks etc. All very thrilling, stopped for a meal under the mangroves. Some rice fields where women gathering rice. On the way back at 1245 a.m. adult female osprey on a dead tree flew up a small side creek and a little later saw it flying back out towards the main river.

Cape Verdes

The Cape Verde islands lie off the western coast of Senegal, 400 kilometres out in the Atlantic Ocean. The eastern islands are low, sandy and dry, but the massive rugged cliffs of those in the north-west group rise high out of the sea. The islands have various endemic species, and a particularly interesting group of raptors including two species of island kestrels, buzzard, peregrine falcon, kites and ospreys. Sabine Hille, who carried out research on the kestrels for her PhD and became interested in the few remaining kites, asked me to come and have a look at small numbers on Santa Antao, which she thought were the distinctive Cape Verdes kite; it gave me a chance to look at the ospreys there.

Santo Antao, Cape Verdes

Sadly, the special kites of Santa Antao are now gone, but there are still about 70–80 pairs of ospreys on the Cape Verdes. They are very isolated; at present, it is thought that they do not leave the island, but I think the population is occasionally topped up by lost vagrants from northern Europe. There is now one recovery of a Scottish youngster, ringed by Roger Broad. These ospreys fish in the sea and have occasionally nested on palm trees, but most of the nests are either on the ground on remote islands or in high cliffs and extinct volcanoes. Life is hard for some of the islanders so eggs and young have been eaten in the past and may still be.

Sao Vicente. Monday, 17th February 1996. South coast. Two adult female ospreys mild territorial dispute over the sea 'tchupping' at each other – one to the west to fish in a rough sea and the other flew east, both were dark birds. In the distance, two hump-backed whales breaching and displaying in the ocean, halfway to Santa Lucia, very big floppy front flippers – black-and-white, being held sideways out of the water. Later climbed old volcano to check osprey nest; made of palm leaves, fig branches and scrub and lined with grass, paper, old clothes and courlene fish netting. May have had young last year. Fish remains below roost sites in the cliffs just underneath; lots of osprey droppings, fish remains including the wings of flying fish and teeth of coral eating fish ....

Santa Antao. 23rd February 1996. Driving over steep, trying dusty road to the west coast; pulled off to look down into the big Ribeira de Tarrafal. Fantastic views down into the gorges; small waterfall in cliff face; five kestrels along top of the gorge which looked like a family party. Then heard osprey calling from end of the ridge and could see female osprey in a nest, (eggs or small young) calling for food – incessant hunger calling. Nest on a point of small cliff in a huge gorge, near the top of probably over thousand-metre cliffs, about two kilometres from the sea – perching rocks with droppings close to it. Incredible place for ospreys to nest. Safe from humans. Very, very hot.

Small numbers of ospreys breed along the north coast of Africa, as well as on the Atlantic islands south to Cape Verdes, and another small population breeds along the Red Sea coast and as far south as Somalia and Eritrea. These birds are linked with those in Saudi Arabia, as proved in January 2001 by Jugal Tiwari when carrying out bird surveys of Green Island, near Massawa. He identified that one bird of a pair of ground nesting ospreys was ringed. It was later identified as a juvenile from the Farasan archipelago.

Alan Poole at the New England salt marshes

North America

The United States

My first visit was to New England where I met Alan Poole, who was carrying out research on ospreys in the tidal salt marshes of Connecticut and New Massachusetts. I was on my way to an osprey conference in Montréal and Alan offered not only to drive me from Boston to the conference but also to show me his study area. In late October the ospreys were, of course, far away in sunnier climes, but I was able to see nests in trees on small islands as well as the nesting platforms which had been established all over the salt marshes. The nests were low and there were few available perches, so Alan had devised special ones which weighed his study birds throughout the nesting season, and weighed the fish they caught and how much they ate. It was a really interesting day and encouraged me to think of new projects with our own ospreys. Some years later, Alan visited me in Scotland as part of his research when writing his important monograph, *Ospreys – a Natural and Unnatural History*.

North American ospreys suffered serious declines in the DDT years of the 1950s and 1960s, leading ultimately to their extinction or great scarcity in many areas. Following the banning of DDT spraying over marshes and wetlands, ospreys in North America started to recover. They had help in reclaiming their lost lands from a succession of reintroduction and translocation projects, as well as provision of artificial nests, and the osprey is, nowadays, well distributed throughout the country. They breed from Florida to Maine, from California to Washington, and are well distributed through the northern inland states as well, with a possible population of 16,000–20,000 pairs; another 10,000–12,000 pairs breed in Alaska and Canada, and 800 pairs in Mexico, giving a continent total in the order of 30,000 pairs. Most of these ospreys winter in Central and South America, moving south to northern Argentina and Chile.

I have been back to the States on several occasions and have always been struck by the tameness of the ospreys there, and by their ability to live alongside people in busy places, nesting on all sorts of structures. I am told that up to 90% of nests in some areas are platforms provided for ospreys. The shooting of ospreys was never a feature of North America's history, and neither was egg collecting, so the birds had – and indeed have – little reason to fear man. I am sure that in the future ospreys will become tamer and more used to humans in Europe, encouraging them to breed in many busy places, as they do now in the States.

Chesapeake Bay. 9th March 1994. Canada goose study visit. Drove to Terrapin beach: saw an adult bald eagle perched in dead tree and northern harrier flying over the marshes. Walked to the shore through wildlife area with muskrats and phragmites reeds;

shoreline rather grubby, picked up two dead long-tailed ducks and a goldeneye. Looked across the bay to Bay Bridge to a new osprey nest platform in the salt marshes with an osprey circling overhead. An early bird back ...

In 1996, while we were working on proposals for translocating ospreys to Rutland Water, we were greatly helped by the experience gained in North America, particularly from Mark Martell in Minneapolis, who was carrying out a project at that time. Tim Appleton, Hugh Dixon and I went out to meet him in order to learn as much as we could for our own project: it was a most successful visit.

Northeast into Wisconsin to Crex Meadows. 27th April 1996. Brilliant waterfowl area, stopped at Fish Lake where an osprey was fishing – also turkey vultures, bald eagles and red-tailed hawks. Saw that three out of the six osprey nests were occupied by a bird standing on the nest, just arrived, and also one bird hunting. Saw two pairs of bald eagles, at least four pairs of northern harriers, merlin, two Cooper's hawks and American kestrels. Had superb views of sandhill cranes, lots of display and trumpeting, and also three pairs of trumpeter swans and a whole host of ducks including buffleheads and hooded mergansers. Several beaver dams, a couple of muskrats, lot of whitetail deer and brilliant views of the female black bear out into the marsh from the woods with three small black cubs, one hitching a lift on her back ...

28th April, 1996. To State Carrer Park to the west of Minneapolis to look at the osprey hacking sites and nesting platforms. Saw several pairs of ospreys floating about and two females on nests – plus an extra pair chasing them around. Walked to another

Osprey nesting in heronry in New Hampshire (photo courtesy of Iain McLeod)

eyrie overlooking a lake: lots of ducks, trumpeter swans, Forster's and Caspian terns, and Bonaparte's gulls, a pair of ospreys on nest pole – the male collecting water weeds for nest building. The female on nest was Mark's satellite-tracked bird just back from South America, just possible to see the radio on the back of the bird.

In 1999, when we first wanted to use satellite transmitters to study the migrations of ospreys from the UK to Africa, we received help again from our North American friends. The following year I was invited to join them when they caught Florida ospreys for their migration studies. It was a place I had always wanted to visit because I knew how tame ospreys were in that state.

Florida. 15th March 2000. With Mark Martell and Matt Solinsky to fit satellite radios to Florida ospreys. Met Mike McMillian and Bud Tordoff and assistant. Rigged up boat trailer and drove 50 miles through orange groves to Lake Istapoga. Launched boat at Cypress Island – great white and blue herons and lots and lots of ospreys. Set off along the south shore – 300-year-old cypress trees growing in a band along the edge of lake, many hanging with long moss. Osprey eyries really close in places, for example, 10 nests in quarter of a mile and as close together as 20 metres. In fact like a rookery! Looking through nests with brooding females and caught one female with a noose mat on nest. Mark fitted satellite radio and we tried to catch her mate but no luck. Caught several other females but in following days tried to catch males as well using my fish trap idea. We had birds attacking it and but no luck and finally on 17th March we tried catching a male with a good sized catfish. Watched barred owls in the cypress while waiting for an osprey to come to the fish, when suddenly a seven foot alligator glided forwards, dived and then swallowed the catfish and snapped our black twine. We decided to give up with this idea. Works in Scotland where no alligators!

Sanibel Island. 18th March 2000. With Bird Westall, International Osprey Foundation, to the south end of the island checking all sorts of Osprey nests – some too high, some in dead trees – but on the shore facing the mainland an excellent nest in a garden with small chicks and also an osprey nest low on a boat pier, with male perched on a nearby boat. Many brown pelicans, spotted sandpipers etc. Put the nest trap on the garden nest and caught the female quite quickly on small young and in less than ten minutes caught the male as well. Female went straight back onto the nest carrying her satellite radio and then we released the male. Excellent pair to catch for migration studies.

Sanibel Island male

I later heard from Mark that the satellite birds threw up some unexpected results: not only did a bird from Maine winter in Florida instead of in South America, but some of the Florida birds, which were thought to be sedentary, migrated in winter to South America. So much has been learned in the last decade about the migration of ospreys that has added to the traditional study of the visible migration of raptors.

Hawk Mountain, Pennsylvania. 11th September 2003. As the weather cleared, the hawks started to migrate south, mainly broad-winged hawks coming through in small numbers, low at first and then kettles of them spiralling into the sky and heading south. A few sharp-shinned hawks passing and at least two Cooper's hawks and a few bald eagles, as well as local red tailed hawks and turkey vultures and three black vultures. By late afternoon several hundred broad-winged hawks had gone by and also two single ospreys heading south to South America.

Canada

A large population of ospreys lives in Canada, stretching from the coastal salt marshes of Nova Scotia all the way across to British Columbia and far north to the Arctic Circle. All of Canada's ospreys migrate and head south to winter in South America. It is thought that there are about 10,000–12,000 pairs of ospreys in Canada and Alaska. One summer I spent nearly a month in the Inuvik Region and Yukon Territory with wolf and bear biologists, Christoph Promberger and Wolf Schroeder.

Yukon, 2nd August 1995. Canoeing down the Bell River – a real wilderness with grizzly bears. Pair of bald eagles on the side river, both perched in trees, their nest in a tall spruce, a massive bundle of sticks on top of the tree; a bit like an osprey nest. Further down river a female merlin flew across and also a red-tailed hawk calling from a tree. Spotted sandpipers on the edge of the river and several myrtle warblers. So on down the river to the hills where we camped, three broods of American wigeon with eleven ducklings between them. And then suddenly a female osprey perched on her nest, 70 feet up in a white spruce, a big nest on the east side of the river below the hillsides. This pair is well above the Arctic Circle and must be one of the most northerly breeding ospreys in North America. Carried on down the river seeing more wigeon, gray jays and spotted sandpipers and then suddenly as we canoed round a bend, four half grown wolf pups on a sandy flat. They slunk off into the scrub. We climbed ashore and howled and about three adult wolves called back to us.

The Middle East

Ospreys nowadays nest on the coast and islands of the Red Sea and the Arabian Sea, but in the past must have had a wider distribution. They feed on sea fish, including coral-eating species, and many of the nests are on the ground on rocks and debris, or on old lighthouses and buildings on small offshore islands. It is a dispersed population which has suffered from disturbance, persecution and probably also from oil pollution and is believed to number between 400 and 1000 pairs. At the end of the Gulf War I was asked by the RSPB and Birdlife International to visit Saudi Arabia, with Burr Heneman from California, to assess the effects of oil spills on bird life in the Arabian Sea and to advise on the clean up and conservation work. It was a very interesting task and we visited many parts of the country.

10th March 1991. Managed at last to get out to the islands by RAF helicopter to carry out oiled bird surveys in the Arabian Sea. Landed Jaryd island, walked right round and counted birds and turtle sites – no oil except remains of the 1987 spill. Osprey with fish, female, and old osprey nest about a metre high on ground, 20 metres from the tide line and two metres across: no recent occupation signs. And an even older nest near it.

Superb view of a bimaculated lark. Landed at Karayn island at 1020, walked round the island – Caspian plover and a few oiled Socotra cormorants, 120 swift terns ... then on to big island where saw an osprey nest with young perched on top of a 16 feet high navigation tower. Nest about a metre high and the adult was flying about ...

Asia

Ospreys breed across northern Russia and Japan, and winter south in the Philippines and Indonesia. Several winter visits to Java and Borneo went by without my seeing a single osprey, but I had more luck in Japan. My first two visits to Hokkaido were to study Steller's sea eagle in winter and to help their conservation. At the time I was simply shown an island on a lake where they nested in summer. My host, Keisuke Saito, though, did take me to the south of Japan to visit the crane wintering grounds and there I did see some wintering ospreys.

Southern Japan. 22nd February 1999. At the Arasaki crane reserve to watch the early morning feeding of about 9000 wintering cranes from Mongolia and Siberia, mainly hooded but also several thousand white-naped cranes, also single sandhill and common cranes. Later went to the coast to look for spoonbills and from the sea wall could see nine ospreys perched on tall bamboos across the bay. The bamboo poles supported long stretches of nets, apparently to protect the seaweed and/or fish farms from bird predators – there were lots of dead birds in the nets – identified mainly as pintail and wigeon ducks but also cormorants, egrets and gulls. Four ospreys were eating fish and plumage-wise they look like ours. But they must be at great risk from the netting!

Eyrie on the coast of Hokkaido, Japan

A few years later I returned to Hokkaido in summer and learnt a little more about the ospreys. The birds were nesting in the most dramatic sea cliffs; but that summer some of the nest sites we visited were empty. I just wonder how safe the birds are, from both persecution and accidental mortality, while on migration south in winter.

Hokkaido, north Japan. 15th June 2003. With Keisuke and Yukiko drove southwest out of Sapporo to the coast. Saw two male ospreys searching for fish in a small estuarine river mouth near atomic power station – saw one stoop but no successful dive. On dramatic headland to the north was an osprey nest on an amazing volcanic pinnacle overlooking the sea, couldn't see nest properly and no sign of birds. Shown where ospreys have put some sticks on high electricity pylon. Then drove south to other coast and saw another osprey nest on top of volcanic pink pinnacle rather like a Corsican nest site, but not occupied ...

I returned to Japan in November 2004 to lecture at an osprey conference in Tokyo with Pertti Saurola from Finland, and, first, to give advice on osprey conservation. Our host Dr Abe took us on a pre-conference tour to northern Japan to discuss osprey conservation and to see some of their breeding sites.

4th November 2004. Driving north along the coast past Sendai – nice river mouth and open sea, north of the big port. One osprey fishing in sea, with shallow dives and then saw three different ospreys perched on posts in the estuary. Could see with the telescope they were all adult males and the closest was not ringed. A nice place to try to catch them.

Australia

My most far-flung ospreys have been those in Australia, where there is a distinct subspecies nesting in three different regions of the country. A pretty strong population nests from northern Queensland south along the coast to Sydney; while another lives in Western Australia and the third, a quite small population, is isolated in southern Australia. When I visited my son and daughter, who both lived in Australia with their families at that time, I was able to see the birds for myself, first in southern Australia and, two years later, in the north.

Kangaroo Island, Australia. 26th November 2002. Visited nest on coast with Rona, Emily and Eric. Female on the nest with young on small rock stack on tide line. She alarm-called at 80 metres and flew off and then back. Well streaked breast and a very white, flat looking head; different to ours. Thin black ring on the left leg and a metal ring on the right leg. Another female came by with half a fish. Female's intruder alarm calling with low call 'cher –chip' not like ospreys at all but her alarm call at our presence was similar to our birds. Nest about three metres across and a metre high; easily reached especially with small ladder. Sanderlings on shore and also sand goanna. Later told by Terry Dennis that the nest we saw was sketched by a person from a 1863 shipwreck and known as the hawks' nest. This nest was unsuccessful until 1987 due to disturbance, but the road line was changed and then the birds were successful. Terry has been studying ospreys for over 17 years and he mentioned identifying the individual adults by the black flecks in their eyes.

Australian pair and nest, Kangaroo Island (photo courtesy of Terry Dennis)

Cairns. October 19th 2004. With Rona and Emily (daughter and grand-daughter) to the park near their house and to look at the osprey nest on a radio tower amongst the houses. The female osprey was standing on her nest and the young lying down. Little cuckoo shrikes in the park; hurried home before a rainstorm and lightning. On subsequent days saw ospreys feeding in the bay at Cairns and on a trip to the south of Cairns three osprey nests on high power pylons – one at Hill Road with one large young, another two further south. Later saw another couple nests on power pylons near Marreba.

Of course, wherever ospreys occur in the world, each population will face its own problems. I recently lectured in New England, where ospreys nest in close proximity to man: on the floodlights over a football pitch, on low-level platforms only yards from private homes and on pylons alongside busy main roads. There the audience could barely believe that egg collecting had ever been a hobby here in Britain, let alone that it remained a threat to our breeding birds. In West Africa, where Scottish birds spend their winters, the fish population on which they rely for food is being rapidly eroded by European fleets, who have acquired rights in African waters. And, in a few parts of mainland Europe, ospreys are still shot, either for sport or because of their natural interest in commercial fish farms. The challenge to all who want to see this magnificent bird restored to its full range around the world is to work with local people, persuading them, bit by bit, that man and bird can peacefully co-exist, and showing them how this can best be achieved. It is also important to recognise that protecting ospreys in their winter quarters or on migration is as essential as protecting their nests.

A favourite roost

12

Where and how to see ospreys

Enjoying ospreys

Ospreys are easy to identify and great to watch. Anyone who has seen an osprey dive into a loch or lake to catch a fish, watching it struggle to rise out of the water and shake itself before flying off, will always remember it as a special moment. But how do you see an osprey if you don't know where to look? Nowadays it is relatively easy because there are a range of public sites where you can go to watch ospreys without disturbing them. You can take your own binoculars or telescope or use the high-powered equipment available in the visitor centre. Most of the sites also have a closed-circuit television link to the nest tree, so you can watch the birds close up. There will be staff on hand who can provide information about ospreys and often there is written material and displays available for you to read. Detailed below is a list of public viewing sites, but remember that these are wild birds and may not be present when you visit, especially if they have failed to breed.

It is also, of course, possible to see ospreys in many other places in the British Isles. My favourite activity is to watch ospreys hunting fish and plunging spectacularly into the water. There are one or two places, such as the Rothiemurchus fishery near Aviemore, where you can be pretty certain of seeing the birds but there are many other waters where you can wait and watch and you may be lucky. If you see an osprey quartering the water, sit down and keep on watching. It can take anything up to half an hour and four or five attempts before an osprey is successful in catching a fish. Maybe it will then just go to a loch side tree and start to eat,

Findhorn Bay

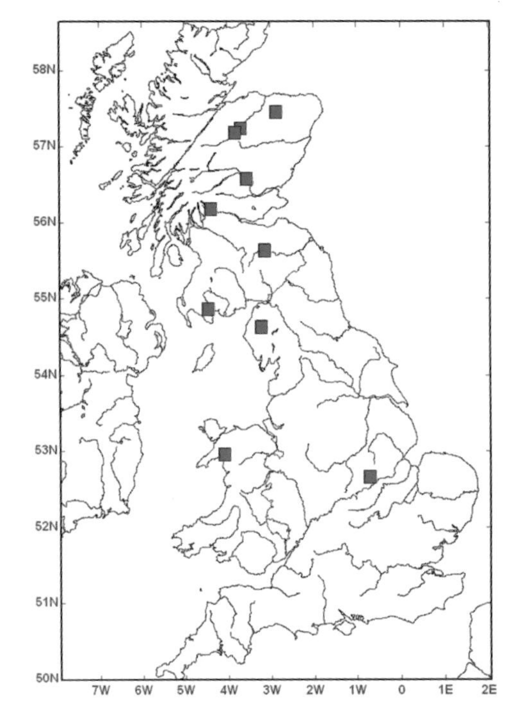

Visitor centres in the UK

rewarding you with great views. One of my favourite places to watch ospreys fishing is where the River Spey empties out into the Moray Firth. Here, in the small estuary of the Spey, ospreys are regularly to be seen fishing; close by is a bronze statue of an osprey holding a fish and the Whale and Dolphin visitor centre. Another is Findhorn Bay, just to the west, where you can watch the ospreys hunt flounders while enjoying a beer at the Kimberley Inn in the evening.

Loch Garten, near Boat of Garten, Strathspey

(OS Grid Reference: NH977183)

The world famous RSPB Osprey Centre is signposted from the A9 and the nearby villages. A large car park and visitor reception centre with eco-toilets is located by the loch. The Osprey Centre overlooks the famous Loch Garten nesting site, the ancestral home of the osprey in Scotland. Binoculars and telescopes are provided for viewing the birds directly or you can watch the close-up action at the nest via live CCTV camera link. Members of staff are on hand to explain the action as it happens. To help you time your visit, the following may be useful:

- early/mid April: the ospreys arrive, settle and lay eggs, and incubation begins;
- late May/early June: hatching takes place and you can see the adults feeding the young ospreys;
- mid-June to mid-July: the young grow rapidly on their diet of fish and are clearly visible on the nest;
- mid-July to mid-August: the young fly for the first time, but remain around the nest for another month;

- from mid-August: the young begin to spend time away from the nest, developing their fishing and flying skills before departing for Africa.

Loch Garten

Capercaillie viewing (Caper-watch) is available also from the centre at certain times of the year – 05.30 a.m. to 08.00 a.m. during April and early May only – with CCTV cameras trained on lekking birds. Additional nest-box cameras are linked to the centre enabling privileged views of box-nesting species such as blue and great tits, redstart and goldeneye duck.

- The Osprey Centre
  Opening hours: 10.00 a.m. to 6.00 p.m. from April to the end of August. Admission charges: Adults £3, Concessions £2, Children 50p, Family Ticket £6.
  Tel: +44 (0)1479 831476 (April to August, 10 a.m. to 6 p.m. only)

- Abernethy Forest Reserve office
  Tel: +44 (0)1479 821409
  Website: www.rspb.org.uk

Latest nest and camera at Loch Garten

Rothiemurchus Fishery, Rothiemurchus, near Aviemore, Strathspey

The fishery car park and shop is close to Aviemore on the road to Cairngorm.

This site offers a unique opportunity to watch the magnificent hunting osprey in action. The best time to visit is in the early morning or evening, when as many as four or five fishing ospreys can be present, but others can visit throughout the day. Visit the fishery and watch the ospreys as they dive to catch rainbow trout from the lochs. There is a bird hide with information displayed about ospreys. The estate encourages visitors and has the most amusing sign at the gates of the fishery – 'osprey now fishing'– always put up once the birds have come back from Africa. The Rothiemurchus Fishery is open every day to allow access to the newly upgraded bird hide. Special arrangements are possible for photographers to hire special hides outside normal opening hours, for example, in the early morning. Prior booking is necessary and can be made by contacting the fishery's office (details below).

- Rothiemurchus Fishery
  (OS Grid Reference: NH897115)
  Opening hours: April–mid-May 9.00 a.m. to 5.30 p.m.; mid-May–August 9.00 a.m. to 9.00 p.m.; September–October 9.00 a.m. to 5.00 p.m.
  Admission charges: Adults £1.50, Children & Concessions £1.00.
  Tel: +44 (0)1479 810703
  Email: info@rothie.co.uk
  Website: www.rothiemurchus.net

Huntly Peregrine Watch

The Peregrine Wild Watch at the Forestry Commission viewing point, just west of Huntly on the A96 trunk road between Aberdeen and Inverness, offers excellent views of a pair of breeding peregrines in a quarry and also available from 2007, CCTV coverage of a nesting pair of ospreys in the surrounding forests.

- Huntly Peregrine Watch
  (OS Grid Reference: NJ497428)
  Tel: +44 (0)1466 760790

Loch of the Lowes, near Dunkeld, Perthshire

The Scottish Wildlife Trust Visitor Centre and Wildlife Reserve is located off the A923 (follow the Wildlife Reserve signs). The nearest train station is Birnam and Dunkeld. The reserve is part of the Fungarth walk path network. The Loch of the Lowes is a beautiful reserve in an extremely scenic location on the Highland–Lowland boundary. Well over one million visitors have watched the ospreys here and these birds have produced 63 healthy chicks between 1971 and 2007. The hides are less than 200 metres from the eyrie, which is situated on a large Scots Pine in the woodland fringes of the loch. The view from the hide is across a small bay. With the nest, the hide and the loch all in one location, it is not uncommon for visitors to see the birds catch a fish right in front of their eyes! Apart from the ospreys, which are the star attraction, visitors can also see great-crested grebes, goldeneye and other waterfowl, as well as red squirrels. The visitor centre is open from 30 March to 30 September from 10.00 a.m. to 5.00 p.m., while one hide is open 24 hours a day, all year round. It is situated close to the A9, which is Scotland's main traffic route north. The car park can accommodate coaches on request.

The hides and centre are easily accessible for people of all abilities and ages. Toilets and hides have disabled access and disabled parking is also available. The reserve and centre now employ up to six staff during the summer season, who are assisted by over 90 local volunteers. CCTV cameras and a webcam on the Scottish Wildlife Trust's own website offer the public, both on and off site, a close-up view of the nest. Telescopes and binoculars are provided in the hides and the centre also features interactive displays and a wildlife gift shop, and light snacks are available when refreshment is needed after some dedicated osprey watching.

- Loch of the Lowes
  (OS Grid Reference: NO050440)
  Admission charges: Adults £3.00, Concessions £2, under-16s free of charge, SWT members free of charge.
  Tel: +44 (0)1350 727337
  Website: www.swt.org.uk

The Trossachs

Queen Elizabeth Forest Park, Achray Forest
(OS Grid Reference: NN520014)

The visitor centre, formerly the David Marshall Lodge, is situated one mile north of the village of Aberfoyle (Dukes Pass) – the 'gateway' to the Trossachs – on the A821. The centre features a new wildlife room, which is manned by a team of volunteers, and cameras are set up to capture the arrival of the ospreys and monitor their nesting and the rearing of their young. The visitor centre offers a 'hands on' experience with minimal disruption to the birds. A ranger is also available to answer visitors' questions. The centre is also the ideal place to start when exploring the rest of the Queen Elizabeth Forest Park.

> *'Live pictures beamed from the nest can be viewed in a room similar to a field station.... In addition to the ospreys it is hoped to obtain pictures of buzzards, blue tits, ravens and peregrine falcons.*
>
> *'So if you are not lucky enough to see these wonderful birds in the woods come up to the Lodge and see them on live television.'* – Forestry Commission Scotland

- The David Marshall Lodge Visitor Centre
  Opening hours: 10.00 a.m. to 4 p.m.
  Tel: +44 (0)1877 382258
  Website: www.forestry.gov.uk/aberfoyleospreys

Tweed Valley Ospreys

The Tweed Valley Osprey Project (TVOP) was set up as part of an initiative to celebrate the 50th anniversary of the return of the osprey to British shores in 1954. Two new osprey centres at Glentress and Kailzie Gardens were set up near Peebles in the Scottish Borders. Whilst the tenet of the TVOP is to protect nesting ospreys, the initiative also aimed to 'share' the experience without compromising the birds' safety.

In 2004 the first pictures of the nesting birds were beamed into the viewing rooms. The live pictures show the ospreys at a nest in the local forest, however, there is no viewpoint of the nest. In 2008 the TVOP is celebrating the tenth year of ospreys nesting in the Scottish borders and to mark the anniversary a new blog has been set up to relay the progress of the nesting residents.

At the Glentress Osprey Centre there is a dedicated osprey room with lots of information and an osprey warden on site. Set in the Glentress Forest, one of the most diverse forests in Scotland, live footage of the nests can be viewed from the comfort of the centre. Glentress Forest also offers world-class mountain biking facilities and a superb selection of walks. To get to the centre, which is located approximately a mile and a half from Peebles, follow the A72 and watch for signposts to the Osprey car park.

The Kailzie Gardens' Osprey Watch Centre is located approximately two and a half miles east of Peebles on the B7062. Ospreys can be viewed by the TV cameras focussed on the nest 30 metres above the forest floor. A great place for the young and young at heart alike, the facilities include an 18-hole putting green, children's play area and well stocked trout ponds, as well as a licensed restaurant, gift shop and, of course, the acclaimed gardens. The osprey centre has proved to be a prime attraction.

Both Centres are open daily from 10.00 a.m. to 5.00 p.m., from May to August. Both offer excellent refreshments and have disabled facilities. For both sites, follow the 'Osprey Watch' signs.

- Glentress Osprey Centre
  (OS Grid Reference: NT283397)
  Admission charges: car parking £2.00
  Tel: 01750 721120

- Kailzie Gardens
  Admission charges: Osprey Watch Centre £1.00 (children free of charge), other charges may apply
  Tel: 01721 720007
  Website: www.tweedvalleyospreys.co.uk
  For further information call the Osprey Information Line on +44 (0)845 FORESTS (3673787) or visit Peebles' Tourist Information Centre.

Wigtown

The 'Ospreys are back in Galloway' project involves Scottish Natural Heritage, the RSPB, D&G Constabulary and local landowners and volunteers, and further heralds the return of the osprey to Scotland. Wigtown is six miles south of Newton Stewart in Dumfries and Galloway. Upon arrival in Wigtown just look for the largest building in the central square; the camera room is beside the Wigtown Bay LNR Visitors' Room on the top floor. Dumfries and Galloway Council had the camera room installed in the County Buildings to provide 'unparalleled' views of the nesting ospreys. The CCTV camera is positioned only a few feet away from the centre of the nest but the location of the nest is strictly confidential and there are no public viewing facilities in the field. The nesting pair are the first in Galloway for over 100 years.

Researchers have identified the fish in the local area and found species such as flounders, mullet and garfish.

- Wigtown
  Opening hours (camera room):
  Monday–Saturday, 10.00 a.m. to 5.00 p.m. with extended hours on Tuesday, Wednesday and Friday when it is open until 7.30 p.m. Sundays, 1.00 p.m. to 5.00 p.m.
  Free admission
  Tel/Fax: +44 (0)1988 402673
  Website: www.dgcommunity.net/osprey
  For further information contact:
  Elizabeth Tindal, Countryside Ranger,
  County Buildings, Wigtown DG8 9JH

Trial nest at Wigtown

The Lake District

Bassenthwaite Lake, near Keswick, Cumbria

The Lake District Osprey Project has two centres, Dodd Wood and Whinlatter visitor centre. The project is a partnership between the Forestry Commission and the RSPB and is fully supported by the Lake District National Park Authority, Bassenthwaite Lake being owned and managed by the National Park Authority. It is a favoured fishing area for the ospreys and the birds nest in a tree on Forestry Commission land.

The Dodd Wood viewpoint is an open air viewpoint. At the time of writing work had started on constructing a new viewpoint as the resident pair of ospreys moved nest and no longer can be seen from the existing viewpoint. It does, though, still provide magnificent views of the ospreys fishing from the lake. The existing viewpoint is situated approximately three miles north of Keswick, off the A591 – follow signposts to the Mirehouse from the A66. Car parking is available at the Sawmill Tearoom opposite the entrance to Mirehouse. From the tearoom the viewpoint is a 15–20 minute walk uphill along a gravel path. Over the course of most days the ospreys can be seen fishing, sitting, feeding, flying and washing. High-powered telescopes are provided but is advisable to bring your own binoculars to gain the very best experience of osprey watching. It is also worth noting that on bright sunny days the visibility in the afternoons can be significantly poorer than in the mornings.

The Whinlatter Visitor Centre is located west of Braithwaite on the B5292 between Braithwaite and Cockermouth and features the indoor osprey exhibition with giant, live video screen to which images are beamed from the cameras fixed on the nest. Project staff are on hand to answer questions and provide a guide to what can be seen on the live video. A slight, and only very temporary, glitch is that the cameras are, at the time of writing, fixed to the old nest, which has now been abandoned by the osprey pair in favour of another.

At both the visitor centre and the open air viewing point, the dedicated and friendly staff are on hand from 10 a.m. to 5 p.m., seven days a week from April until the end of August, to answer all queries and report on the progress of the nesting ospreys.

- Dodd Wood viewpoint (OS Grid Reference: NY235282)
- Whinlatter Visitor Centre (OS Grid Reference: NY208245)
  The exhibition is open from 10 a.m. until 5 p.m. seven days a week (usually April until the end of August). There is no charge for the viewpoint or exhibition, though, car parking charges do apply.
  Tel: +44 (0)17687 78469
  Website: www.ospreywatch.co.uk

Viewing point at the Lake District

Rutland Water, Oakham, Rutland
(OS Grid Reference: SK879072)
Rutland Water nature reserve is the home of the first osprey translocation project in Europe; ospreys now breed here and fishing birds can often be seen from the hides on the reserve. This is a most superb bird watching area with two visitor centres, 15 birdwatching hides and lots of trails.

The Anglian Water Birdwatching Centre, on the Egleton section of the reserve, lies south-east of Oakham, the county town of Rutland. From the west, the reserve is reached via the A6003 from Uppingham or Oakham. Coming from the east, use the road from Stamford – A606/A1. Or why not come by train? Oakham station is only two miles away and is served by an excellent east–west service linking Peterborough to Leicester and Birmingham.

The centre has an excellent display on ospreys and knowledgeable staff and volunteers on hand daily. The Osprey staff members provide talks on the birds, and sometimes lead walks and Osprey cruises on the reservoir. A great variety of species other than ospreys may be seen here, and it is also the home of the international British Birdwatching Fair, which takes place in August. There is a gift shop, optics shop and toilets. The osprey project officer is Tim Mackrill. During the summer months there is an extensive display about the Osprey Project, and video footage from the osprey nest.

The Lyndon Reserve visitor centre is signposted from Rutland Water South Shore, one mile east of Manton. Since 2007, a pair of ospreys breed on a nesting pole in the water, which is clearly visible from several hides at the Lyndon visitor centre. The presence of the nesting pair has attracted thousands of visitors.

- Egleton Reserve
  (OS Grid Reference: SK879072)
  Opening times: Open daily, throughout the year from 9 a.m. to 5 p.m. (4.00 p.m., November–January).
  Admission charges: Free parking.

- Lyndon Reserve
  (OS Grid Reference: SK894058)
  Opening hours: from 1 April to the end of October, Tuesday–Sunday.
  Free parking.
  Tel: 01572 770651

  Permits are needed to visit both sites and are available from the Anglian Water Birdwatching Visitor Centre, Egleton
  Websites: www.rutlandwater.co.uk, www.ospreys.org.uk

Glaslyn Osprey Centre, Pont Croesor, Portmadog, North Wales
A pair of ospreys nested for the first time in 2004 and in co-operation with the landowners, North Wales Police, Countryside Council for Wales and local volunteers, RSPB Cymru set up a viewpoint at Pont Croesor, which is staffed daily. The ospreys can be viewed from the hide on large plasma screens via CCTV footage, and also by binoculars and telescopes. There is only one viewing point available, and assuming that the ospreys return, the view point is set up off the B4410. The views of the landscape of Snowdonia are also impressive.

It is advised that visitors consider using public transport to visit the Glaslyn visitor centre. To travel by car, take the A487 north from Porthmadog and turn right onto the A498 at Tremadog. Turn right onto the B4410 after approximately two and a half miles. It does help to avoid congestion on this minor road by making your return journey on the B4410 to Gerreg and on to the A4085. Since 2004, thousands of people have visited the Glaslyn Osprey Project.

- Glaslyn Osprey Centre (OS Grid Reference: SH594414)
  Opening hours: daily from 10 a.m. to 6 p.m.
  Website: www.rspb.org.uk/brilliant/sites/glaslyn

† Please do contact individual centres directly for all up-to-date information, including opening times, when planning a visit.

Appendix: Osprey miscellany

Common name

English – *Osprey*
Danish – *Fiskeorn*
Dutch – *Visarend*
Estonian – *Kalakotkas*
Finnish – *Sääksi*
French – *Balbuzard pêcheur*
Gaelic – *Iasgair*
German – *Fischadler*
Hungarian – *Halászsas*
Icelandic – *Gjodur*
Hindi – *Măchhmănga*
Italian – *Falco pescatore*
Japanese – *Misago*
Norwegian – *Fiskeorn*
Polish – *Rybolów*
Portuguese – *Águia-pesqueira*
Russian – *Ckona*
Spanish – *Águila pescadora*
Swedish – *Fiskgjuse*
Welsh *Gwalch y Pysgod*

Scientific name
Pandion haliaetus

Distribution of breeding ospreys in Europe

Subspecies

Races or subspecies
P.h.carolinensis: North America
P.h.cristatus: Australia, South-east Asia
P.h.haliaetus: Europe, Asia, North Africa
P.h.ridgwayi: Caribbean

Palaearctic Osprey – *Pandion haliaetus haliaetus*
This is the main or nominate subspecies of osprey in Europe and Asia and it breeds from the United Kingdom eastwards, across Europe and Asia to Japan. The breeding range extends from just north of the Arctic Circle in Finland to as far south as the Cape Verde islands, North Africa and the Red Sea while in the east, it nests from Kamchatka to the southernmost islands of Japan. Most of these populations are migratory and spend the winter in Africa, India and southeast Asia. The bird is a large, dark breasted osprey, with a well marked head.

North American Osprey – *Pandion haliaetus carolinensis*
This race breeds from Alaska across northern Canada, just reaching the Arctic Circle, to Labrador and southwards to Florida in the east and Baja California in the west. In winter the birds migrate to South and Central America, with most found in the northern half of South America. They are smaller and paler ospreys than the Eurasian species and not as large.

Caribbean Osprey – *Pandion haliaetus ridgwayi*
This is the rarest of the subspecies, probably numbering less than 200 pairs. It is found in Cuba, the southern Bahamas and the east coast of the Yucatan Peninsula; it is thought to be non-migratory. It is paler and slightly smaller than carolinensis and the head lacks a solid band.

Australasian Osprey – *Pandion haliaetus leucocephalus*
This race breeds in Australia, mainly in the coastal districts, and to the north in the eastern islands of Indonesia and New Guinea and on other islands in the southwest Pacific. Most of these populations are thought to be non migratory. It is the smallest race of osprey with a white head with some brown streaks and a breast band that is well developed. Visually, this is the most distinctive of the various races.

Characteristics

Breeding
Age at first breeding
average, 3–7 years
(very rarely, 2 years)

Clutch size
3, sometimes 2, very rarely 4
Single-brooded

Egg size
62 × 46 mm

Fresh weight of egg
72 g

Incubation period
35–37 days
Both of the pair incubate, but mainly a duty of the female

Fledging period
51–56 days

Plumage characteristics
First down plumage at hatching
Second down from 11 days

First complete feathers at 42 days
Full adult plumage at 18 months

Body size (average for adult haliaetus):

	male	female
Length (cm)	56–60	57–62
Wingspan (cm)	147–166	154–175
Tail (cm)	19–21	20–23
Weight (kg)	1.4–1.6	1.6–2.2

The arts and media

Photography and film

Over the years there have been some excellent films made about ospreys. The RSPB produced a film *The Return of the Osprey* in the early 1970s and this was later followed by the spectacular, groundbreaking production *Osprey* filmed by Hugh Miles, which was shown by the BBC and released worldwide. The RSPB have documentaries and films about ospreys on sale. A new film about the Rutland Water project was released in 2005; this is available from Rutland Water Nature Reserve.

Several of Britain's finest photographers, including Laurie Campbell, Chris Gomersall and David Whitaker, have worked with ospreys and some of them have their work displayed on their own websites. Photographs can be ordered from them.

Art and sculpture

The osprey lends itself to being depicted by means of painting, drawing, sculpture, jewellery and other art forms, and one can find beautiful examples of fine work that takes the osprey as its inspiration. One of the most famous artists to feature the osprey is Keith Brockie, who is well known for his bird studies. He has painted, protected and studied ospreys in Perthshire for many years. There are sculptures of ospreys in bronze at Spey Bay, a wood at Rutland Water and several other localities.

Logos and brand names

The village of Boat of Garten was the first to use the osprey as its logo and this example was soon followed in Aviemore where a huge osprey sign was erected to adorn the Aviemore Centre when it opened in 1972. As ospreys became big news and were often featured on radio, television and in the newspapers, their name and image became favourite corporate symbols for businesses and was regularly used to promote tourism and a whole range of products. Presumably it was good for business because the name 'Osprey' is very well used nowadays! A huge military helicopter in America is called the Osprey. I always think that it is appropriate, in view of the direction my own life has taken, that before retiring my brother was an engineer on that particular project! So we can both lay claim to having been involved with ospreys.

Osprey emblem, Aviemore 1977

Stamps

Ospreys are very famous birds worldwide and have featured on postage stamps in many countries. In my own collection I have a beautiful stamp from Poland featuring a flying osprey. Mongolia has one of an osprey rising from a lake, and there are others from Germany, Holland, France and India.

Sixty stamps from 55 countries are shown on the following website: www.bird-stamps.org/species/29001.htm

Some osprey stamps from author's collection

Spey Bay Osprey cast in bronze

Recommended reading

The following is a list of books about ospreys. Some are out of print but are available in libraries.

Brown, Leslie. (1976) *Birds of Prey*. Collins.
Includes a chapter on ospreys.

Brown, Philip & Waterston, George. (1962) *The Return of the Osprey*. Collins.
Lots of historical information.

Brown, Philip. (1979) *The Scottish Ospreys*. Heinemann.
Much about Operation Osprey.

Dennis, Roy. (1991) *Ospreys*. Colin Baxter Photography.
A popular account, liberally illustrated.

Harvie-Brown, J. A. & Buckley, T. E. (1895) *A fauna of the Moray Basin*. David Douglas. pp. 71–94.
Historical data on ospreys in Scotland in the 19th century.

Harvie-Brown, J. A. & Macpherson, H. A. (1904) *A Fauna of the North West Highlands and Skye*. David Douglas. pp. 178–206
Historical data, includes sketches and lithographs of nesting sites.

Poole, Alan. (1989) *Ospreys: A Natural and Unnatural History*. Cambridge University Press.
This is the best scientific book on the osprey throughout the world and is a most interesting read.

Ramshaw, David. (2007) *The Lakeland Ospreys*. P3 Publications.
A new account of the ospreys at Bassenthwaite.

Foreign books on ospreys

D'Arcy, Gordon. (1999) *Ireland's Lost Birds*. Dublin, pp. 79–84
Chapter on the history of the osprey in Ireland.

Koivu, Juhani & Saurola, Pertti. (1987) *Sääksi*. Finland.
Detailed account and beautiful photographs of ospreys in Finland (in Finnish).

Mebs & Schmidt. (2006) *Die Greifvögel*.
Chapter on ospreys (in German).

Moll, Karl-Heinz. (1962) *Der Fischadler*.
Out of print classic account of ospreys in Germany.

Poole, Alan *et al.* (2002) *The Birds of North America.*
Detailed account of the North America osprey species. An online version is also available.

Thibault, Bretagnolle & Dominici. (2001) *Le Balbuzard pêcheur en Corse*. Ajaccio.
Detailed account of the Corsican ospreys; research and conservation (in French).

Useful websites

Then are now many excellent websites about ospreys around the world, and also an increasing number of live webcams recording their daily lives. The following lists some of the sites but it is not exhaustive. A search on the internet will retrieve a lot about real ospreys, amongst the plethora of sites that merely use the name of this wonderful bird!

United Kingdom
www.ospreys.org.uk
The Rutland Water Osprey Project website is one of the best and has much information about ospreys, the reintroduction project and journeys of satellite-tracked birds, as well as many links.

www.roydennis.org
The author's Highland Foundation for Wildlife website has detailed information on osprey migrations, and other raptors, as well as sections on osprey nest reporting and nest building. It also includes the migrations and life of Logie, the osprey from Moray.

www.rspb.org.uk
The RSPB's website has a very popular section about ospreys, including live webcam updates and diary entries from Loch Garten and Glaslyn during the osprey season.

www.swt.org.uk
The Scottish Wildlife Trust reserve at Loch of the Lowes has a regular breeding pair of Ospreys and its website is updated frequently with news of the pair's progress.

www.ospreywatch.co.uk
The Lake District Osprey project is described on a Forestry Commission website.

European sites
http://balbuzard.lpo.fr
The French osprey site provides much information – and many images – about the expanding breeding population in France. There are summaries in English.

www.fmnh.helsinki.fi/english/zoology/satelliteospreys
www.fmnh.helsinki.fi/english/zoology/ringing/research/osprey.htm
www.lpl-video.fi/product_fishing%20eagle.htm
These three illustrated Finnish sites provide information on osprey migration and satellite-tracking studies, and have photographs.

www2.nrm.se/rc/osprey.html.en
The Swedish Bird Ringing Centre's website details osprey ringing recoveries.

www.kotkas.ee/ENG/strack.html
The Estonian Eagle Club and Ornithological Society website also provides detailed information on osprey tracking (in English).

http://balbuzards.cfsites.org/
Provides information on ospreys nesting on the rocky coasts of Morocco.

www.arabianwildlife.com/archive/vol3.1/prirsea.htm
Details of non-migratory ospreys breeding around the Red Sea (in English).

North America

www.ospreys.com
International Osprey Foundation provides information about ospreys and photos of the birds in Florida.

www.nhaudubon.org/research/nhosprey.htm
Some excellent material about an osprey project in New Hampshire, USA.

www.friendsofblackwater.org/osprey_cam_blog
Ospreys nesting on Blackwater reserve.

www.birds.cornell.edu/birdhouse/nestboxcam/KYarchive1_2004.html
Webcam images and information about ospreys.

www.bioweb.uncc.edu/bierregaard/osprey_maps.htm
Rob Bierregaard's excellent site details migrations of satellite-tracked ospreys from eastern USA.